THE

EXTRAORDINARY

LIFE

OF

CHARLES POMEROY STONE

Brigadier General Charles P. Stone with his daughter Hettie. Stone's wife Maria Louisa would die at age thirty-one in February 1863, leaving Stone a widower with a child to raise. (*Library of Congress*)

THE

EXTRAORDINARY
LIFE

OF

CHARLES
POMEROY
STONE

SOLDIER,
SURVEYOR,
PASHA,
ENGINEER

BLAINE LAMB

WESTHOLME
Yardley

Westholme Publishing, LLC
904 Edgewood Road
Yardley, Pennsylvania 19067
Visit our Web site at www.westholmepublishing.com

First Printing December 2015
10 9 8 7 6 5 4 3 2 1
ISBN: 978-1-59416-232-9
Also available as an eBook.

Printed in the United States of America

CONTENTS

Contents

INTRODUCTION

O N the morning of January 28, 1988, in the East Room of the White House, President Ronald Reagan welcomed then–Egyptian president Hosni Mubarak and his wife to the United States. The bulk of President Reagan's public remarks centered around the ongoing quest for peace in the Middle East and the key role played by Egypt in the process. In his conclusion, however, he referenced an extraordinary former American military officer who served in that country over a century earlier. The president declared General Charles Pomeroy Stone to be "a fitting symbol of our enduring relationship." He went on to mention that Stone later designed and constructed the base of the Statue of Liberty, and likened that solid base to the solid friendship between the two countries.

But just who was this "fitting symbol," and what made his life extraordinary? Certainly Charles Pomeroy Stone's early years did not mark him as extraordinary—a comfortable middle-class upbringing, followed by an education at the United States Military Academy. He did well there, but did not finish at the top of his class. His early military career also differed little from the experiences of many of his fellow junior officers. He joined the Ordnance Department of the antebellum army, which promised steady employment, but slow advancement and little excitement. The Mexican War brought a short respite from the routine, and afterward he was handed substantial command responsibilities, although these alone did not make his life extraordinary. Nor did the personal and professional defeats and challenges he encountered before, during, and after the Civil War. Rather it was his steadfastness and resilience in meeting and overcoming those challenges and defeats that lifted his life well above the ordinary.

Over the course of his sixty-two years, Stone developed into one of those archetypal extraordinary characters of the mid-nineteenth century: distinguished veteran of the war with Mexico, scapegoat of the Civil War, failed banker, controversial diplomat, pasha in the court of the khedive of Egypt, and engineer for the Statue of Liberty. Physical testaments to his engineering skill and ingenuity still stand in California and New York. Yet, outside of a single battle and its aftermath, his exploits are not nearly as familiar as those of other characters of the era, such as George Armstrong Custer and Charles "Chinese" Gordon. Perhaps had he died under a shower of Sioux and Cheyenne arrows or Mahdiist spears, instead of in a bed in Flushing, New York, he would not have drifted into postmortem obscurity.

As it stands, the Battle of Ball's Bluff in October 1861 is the only episode in Stone's extraordinary life that has received much more than a passing nod from historians. Russell Beatie discusses Stone's activities in Washington, DC, during and after the secession crisis, as well as the Battle of Ball's Bluff, in the first two volumes of his history of the Army of the Potomac, as does William Marvel in his revisionist history, *Mr. Lincoln Goes to War*. Both portray him as a competent general, faithful to the Union, who falls victim to his own trusting nature and political intrigue on the part of the War Department and the Congressional Joint Committee on the Conduct of the War. The politicians unjustly punish him for the defeat at Ball's Bluff, question his loyalty, and destroy his reputation. Similar opinions of Stone's misfortunes are offered by Byron Farwell, James A. Morgan III, and Kim Holien in their books on the battle, and by T. Harry Williams in his *American Heritage* article, "Investigation, 1862." As a result, he has come down in history as the ultimate scapegoat, wrongly accused and incarcerated, a man whose life was pretty much a cipher both before Ball's Bluff and after his release from prison.

When Stone appears elsewhere in historical texts, it is usually in a supporting role. In his biography of Mexican *caudillo* Ignacio Pesqueira, *Sonoran Strongman*, Professor Rodolfo Acuna describes Stone—during his short but tumultuous time before the Civil War as a surveyor and consul in northern Mexico—as a troublemaker and likely filibusterer. In later years, Stone served almost as long in the Egyptian military as he had in the American army. William B. Hesseltine and Hazel C. Wolf, in their book *The Blue and the Gray on the Nile*, relate a number of incidents from this period of his life, concentrating on his role as leader of the contingent of ex-Civil War officers who went to serve the Khedive of Egypt. A less flattering view is found in John P. Dunn's more recent *Khedive Ismail's Army*, in which he refers to Stone and his colleagues as often incompetent mercenaries, or "neo-mamelukes." Several citations in Patrick Richard Carstens's *Encyclopedia of Egypt During the Reign of the Mehemet Ali Dynasty* deal with Stone's service to Ismail and Tewfik, but more than occasional factual errors require the reader to verify the author's statements. Although Stone produced no published account of his time in Egypt, reminiscences by William Wing Loring, William McEntyre Dye, Charles Chaille-Long, James Morris Morgan, and others who served there under his command present both positive and negative opinions of the general. Chroniclers of the Statue of Liberty usually include brief mentions of Stone's role on that project, although Elizabeth Mitchell in her *Liberty's Torch* offers a more detailed, if not very complimentary, account of his work as the statue's chief engineer.

One likely reason for the lack of a comprehensive look at the life of Charles Pomeroy Stone is that he left no substantial body of letters or journals, at least none which have so far come to light or found their way into research institutions. Evidence indicates that he was a prodigious correspondent who kept meticulous journals. When he was arrested in 1862, reports mention that his letters and diaries were gathered up and sealed pending a court-martial, which never materialized. Later his personal correspondence was returned to him, but what became of this is unknown. Similarly, his daughter recorded that the family's baggage upon departure from their home in Cairo included journals kept by Stone during his time in Egypt, but they, too, are missing. As a result, it has been necessary to piece together Stone's story largely from the collections and accounts of the people who knew him, as well as from published sources such as *The War of the Rebellion: A Compilation of the Official Records of the Union and Confederate Armies*, the *Report of the Joint Committee on the Conduct of the War*, and his own published recollections of events in Washington, DC, during the early months of 1861. Perhaps this introduction to Stone's life and adventures will bring more sources to light.

Despite his relative obscurity, Stone has appeared as a character in twentieth-century popular fiction. The final episode of the 1961 historical drama television series *The Americans*, "The Inquisitors," presented a version of the congressional investigation that followed Ball's Bluff. Historical fiction author John Jakes included a passing mention of Stone as commander of the District of Columbia Militia, and of his undercover work ferreting out disloyalty in early 1861, in his 2000 novel, *On Secret Service*. Stone also appeared in Michael Killian's 2001 Civil War mystery, *A Killing at Ball's Bluff*. The plot revolves around an investigation (not by Congress) into the death of Colonel Edward D. Baker during the battle. Stone is never a suspect, and his role in the whodunit is fairly minor. Nevertheless, Killian did succeed in capturing pretty accurately the general's "erect bearing, perfectly groomed hair and beard, an honest face, and bright, quick, intelligent eyes."

There are several people who deserve my thanks for their assistance in assembling the story of this extraordinary individual. Arizona historian Bert Fireman introduced me to Charles Pomeroy Stone. As a doctoral student at Arizona State University forty years ago in Fireman's Arizona history course, I was required to calendar a roll of microfilm from the National Archives. The film assigned to me covered the returns of American consuls in Guaymas, Mexico, from the 1840s through the 1860s. About midway through I came across correspondence and reports relating to a group of wayward surveyors led by an ex-army officer named Stone, and their troubles with the governor of Sonora, that piqued my interest. Subsequent research papers for Professor Paul Hubbard's Civil War history courses allowed me to delve further into Stone's role in the defense of the District of Columbia during the secession crisis, as well as the Battle at Ball's Bluff. Here I found tantalizing references to him during the Mexican War, his time in Gold Rush California, the part he played in the disastrous Red River campaign of the Civil War, his service to the khedives of Egypt, and the construction of the Statue of Liberty in New York Harbor. I was hooked. As employment and family responsibilities allowed, my research on Stone continued to the point that it was time to put the story together. The staffs of the National Archives, California State Library, and Sacramento Public Library have provided valuable assistance in assembling the material for this book. A friend, Dave Blottie, furnished valuable insights on the West Point experience. Bruce Franklin at Westholme Publishing took excellent care of the many details involved in bringing the manuscript to print. And, of course, this book could not have been completed without the patient support of my wife, Rosemary, who, as she often reminds me, also has lived with the extraordinary Charles P. Stone over these many years.

HILLS OF NEW ENGLAND

TO THE

WEST POINT PLAIN

T HE blue-clad bodies that began washing up along the banks of the Potomac at Washington City during the first week of November 1861 were battered and bloated. They suffered the effects of having been in the river for days. While some Washingtonians had gotten a taste of war by travelling to witness the Battle of Bull Run that July, the appearance of these corpses brought the tragedy of the conflict to their very doorsteps, and they did not like it. There was no mystery about where the bodies had originated. On October 21, Union troops had been routed at an engagement on the bluffs above the Potomac near Leesburg, Virginia. Many were either killed or had drowned while retreating across the river. Just a few months earlier, they had marched smartly through the capital, proud to be among the first to answer President Lincoln's call for volunteers to save the Union. Now they were barely recognizable. Someone had to be held responsible for ordering these brave young boys to the slaughter. Eventually the onus fell on the commander of the federal forces at the battle, a general who, ironically, had been declared a hero that spring for his role in defending Washington from possible seizure by secessionists. He was Charles Pomeroy Stone—whose fate would become synonymous with military justice and politics gone awry, but whose story involved much more than the debacle on the Potomac.

Charles Pomeroy Stone, West Point graduate, distinguished veteran of the war with Mexico, scapegoat of the Civil War, failed banker, disgraced diplomat, and later pasha in the court of the khedive of Egypt, was born on September 30, 1824, in the northwestern Massachusetts town of Greenfield. His forebears had come from Britain to New England in the 1630s. In later years, Charles would take pride in the fact that his ancestors had fought in campaigns against the Indians, French, and, eventually, the English during the War for Independence. His father, Alpheus Fletcher Stone, was a native of Rutland, the geographical center of Massachusetts. In 1799, twenty-one-year-old Alpheus gave up a teaching career and moved to Greenfield to learn medicine in the office of his brother, John Stone. After two years of study, he opened his own practice and was almost immediately faced with a dysentery outbreak, which he treated successfully. Alpheus Stone later developed a specialty in obstetrics and pediatrics, handling over two thousand such cases in the Greenfield area during his career. A well-respected physician, he became a fellow of the Massachusetts Medical Society in 1814. Alpheus Stone married three times. His first wife died in 1807 within a year of their marriage. Harriet Russell, his second wife, whom he married in 1809, bore him three daughters and two sons before her death in 1817. Five years later, Stone married a widow, Fanny Cushing Arms, of nearby Deerfield. She also came from a distinguished New England family, which included Revolutionary War general Benjamin Lincoln and Thomas Cushing, a member of the Continental Congress and later lieutenant (and acting) governor of Massachusetts. With Fanny, Alpheus fathered another three daughters and two sons, one of whom they named Charles Pomeroy.[1]

Charles spent his growing-up years in and around Greenfield and the lush hill country of Franklin County. Located at the confluence of the Deerfield, Green, and Connecticut rivers, the area around the community had originally been inhabited by the Pocumtuck tribe, which was later driven out by the Mohawks. The site of Greenfield then became the eastern terminus of the

Mohawk Trail and an important trading location. English pioneers colonized the area in 1686, as a part of their Deerfield settlement, but in 1753 they established a separate town of Greenfield. Originally a farming community, Greenfield's economy changed with the development of the South Harvey Canal, America's first navigation canal, in 1795. This allowed freight and passenger boats to reach the town via the Connecticut River, increasing its importance as a commercial center and attracting new settlers, such as Alpheus Stone. The nearby falls of the Green River also provided water power for industry, including the Green River Cutlery Works, manufacturer of the Green River Knife, a favorite among trappers, miners, and emigrants in the Far West. Greenfield became the seat of Franklin County in 1811, and during Charles's youth had developed into the economic, political, and social hub of northwestern Massachusetts.

As a boy, Charles P. Stone attended local schools and academies, including the nearby, prestigious Deerfield Academy. By all accounts, he was a smart and ambitious student, with a flair for mathematics. At age fifteen, he applied through his congressman to the United States Military Academy at West Point. Although he must have had his father's permission, it is unclear what motivated the teenager to seek admission to the academy, as opposed to following Alpheus's footsteps into medicine. Most likely it was his family's military heritage, for in his eulogy for Stone, West Point classmate Fitz-John Porter later speculated that "his soldierly qualities seem to have come to him by heredity."[2] While advancement in the nineteenth century army could be slow, a commission did confer a social standing that few other professions could match. In addition, West Point offered a subsidized education at the nation's foremost engineering and technical college. The nation was growing, and trained engineers would be in the vanguard of that growth. Or Charles may simply have had a desire to experience the world outside of Greenfield, which a West Point education and military career could make possible—or it may have been a combination of all of these. Whatever his reasons, however, young Stone's dream of a West Point education would be deferred, as his initial application was turned down. At age fifteen he would have been young for an appointment, although occasional exceptions were made for applicants with political connections, or for those who showed outstanding promise, as in the case of fifteen-year-old George B. McClellan in 1842. Stone reapplied the following year with stronger letters of recommendation. Albany, New York, attorney and West Point graduate Robert E. Temple described him as "intelligent, gentlemanly, & well educated for a lad of his age," while Luther B. Lincoln, the principal of Deerfield Academy, remarked that he was "esteemed within & without the schoolroom as possessing pure & honorable principles."[3] This time Stone's application was successful, and in the early summer of 1841 he left home and found himself on his way to New York's magnificent Hudson River valley and a military career that would take him to locales he never could have dreamed of, far, far away from Massachusetts's green hills.

As the steamboat approached the landing at West Point, young Charles P. Stone must have been awed by the forested highlands that rose abruptly from the Hudson River. Above the river, he could glimpse ruined fortifications that dated from the Revolutionary War, and as he climbed to the plain above the dock, the buildings that comprised the academy came into view—the barracks, mess hall, chapel, new library, and almost new academic building where classes would be held. Here he would mature from a schoolboy to a military leader and an accomplished gentleman of the antebellum era.

But there was much to be done before classes could begin. Stone and his mates had to pass mental and physical entrance examinations. Being a bright lad with a sound educational background, the former posed no problem, and being a doctor's son, neither did the latter. Those who passed the exams were issued their gray uniforms and quickly shuttled into the cramped, spartan barracks. A couple of weeks later, the freshman cadets moved outdoors and pitched camp on the plain, where they received their introduction to military life, which included artillery practice and infantry drills—drills, drills, and more drills. They also had pounded into their heads the drum cadences that would govern when they awakened, dressed, ate, went to class, marched, and seemingly every other activity at the academy. And the new cadets quickly became acquainted with the academy's rigid system of discipline. At the end of August they moved back indoors, into the two buildings that would be their homes for the next four years. More like dormitories than barracks, they housed at least two cadets per room. The cadets found their accommodations sparsely furnished with just tables, chairs, beds, a closet, a mirror, and a washstand. The rooms were heated by small fireplaces and lighted by smoky, pungent whale oil lamps.[4] There was a place for everything a cadet owned or was issued, and everything had to be kept in its place. Rooms were to be maintained in a spotless condition, or else cadets risked getting demerits if inspections turned up any dust or contents out of order.

Demerits were the bane of cadet life. They were handed out to unfortunate recipients for falling asleep while on guard duty; for cooking "hash" in their rooms; for improper wear of the uniform; for being late for class, meals, or chapel; for chewing tobacco while on duty; for failure to observe "lights out" and any number of other infractions of academy regulations, major or minor. Although years earlier Robert E. Lee had achieved legendary stature by passing through West Point without accumulating any demerits, they were a certainty for more mortal cadets, and a total of two hundred could result in expulsion. Dismissal from the academy was almost a foregone conclusion for unlucky miscreants apprehended sneaking off grounds, as many did, to enjoy the liquid refreshments, tasty victuals, and cheerful environs of nearby Benny Haven's Tavern, whose gregarious namesake and proprietor was a veteran of the War of 1812. By the time of Stone's arrival at the academy, Benny's had already become the traditional illicit gathering spot for daring West Pointers. In all probability, however, the young cadet from Greenfield chose not to endanger his future by partaking of Benny's hospitality, or, if he did, he was one of the fortunate ones who did not get caught.[5]

Charles P. Stone arrived at West Point at sixteen years and nine months of age, barely out of boyhood. Of average height and weight, he had dark brown hair, blue-gray eyes, and a prominent, broad forehead. He did not cut an outstanding figure among the seventy-six members of his class; at least, he was not remembered by fellow cadets in their memoirs. Along with most of the other new arrivals, Stone was most likely confused and a little frightened, all the while trying to project an air of nonchalant confidence. His social life heretofore had been largely with family friends, acquaintances, and schoolmates from the hill country of northwestern

Massachusetts. For the next four years, however, he would associate with cadets of all temperaments and from all parts of the nation. During Stone's time at the academy the cadet corps produced a number of young men destined for fame during the Civil War. Upperclassmen included the diminutive but steady Ulysses S. Grant, whom he would have known as "Sam"; William S. "Rosey" Rosecrans, a seemingly all-around perfect cadet; and James Longstreet, whose hulking frame had "military man" written all over it. Among those entering the academy in Stone's second year were the brilliant and ambitious George B. McClellan, the rebellious and fun-loving prankster George Pickett, and the inscrutable, clumsy, but determined Thomas J. Jackson. In comparison with these cadets bound for greatness, Stone fell somewhere in the middle. He was neither painfully shy and aloof, like Jackson, nor the life of the party, like Pickett. Likewise, while not a natural scholar like McClellan or warrior like Longstreet, he was a serious student and a competent soldier. He strove to be more than middling, but never reached the top of his class. Stone became most closely identified with his fellow cadets in the class of 1845, including Fitz-John Porter, appointed from New Hampshire and a member of the famous navy family whose cousins included David Dixon Porter and David Farragut; Barnard Bee, a popular aristocrat from South Carolina; and Edmund Kirby Smith, informally known as "Seminole," after his native state of Florida. They, too, would achieve notoriety of sorts during the Civil War. What influence mingling with these and other future military leaders had on Stone is unclear, but in years to come he would also become a noted soldier, although not in the way he or anyone else who knew him at the academy could have imagined.[6]

During their first year at the academy, Stone and his classmates were officially called "plebes," but upperclassmen referred to them as "things," "beasts," "reptiles," "apes," and other derogatory terms. Upperclassmen, it seems, did not speak in normal tones—they always shouted commands along with taunts and name-calling. In all likelihood, Stone and his freshman classmates had never encountered such mistreatment. Some may have been shocked by it, and others repelled, but by the 1840s it was a West Point tradition, and they endured the abuse, along with practical jokes from the senior cadets—perhaps because they knew that in the future they would be the ones dispensing the insults. The upperclassmen generally made life miserable for the first-years, whose lives already seemed miserable enough due to the strict military discipline and rigorous academic course. Throughout their time at West Point, cadets would engage in regular infantry drill, artillery practice, and equitation. Classroom instruction was based on memorization and recitation, with students called upon to answer questions or present solutions to problems at the chalkboard before their classmates and under the scrutiny of their instructors.[7] First-year cadets had to master mathematics and acquire at least a reading knowledge of French. Mathematics would improve their problem-solving skills and serve as a basis for later training in engineering. French was the language of Napoleon, whose campaigns formed the basis for much of their military education. French military engineering also served as a model for instruction in that subject. The course of study required the cadets to put in a considerable time studying if they wanted to avoid being asked to leave. Cadet William Dutton of the class of 1846 remarked, "We are obliged to be in our rooms about ten hours a day as study hours . . . if one thinks of staying." The first year culminated in June with the annual examination before the Board of Visitors assigned by the secretary of war to inspect the academy and judge the progress of its cadets. In 1842 it was composed of noted civilian scholars and military officers, including noted U.S. Navy commodore Matthew C. Perry. Stone passed this dreaded event, ending his plebe year with a ranking in the "Order of General Merit" (a combination of academic and conduct roll standings), eighth in his class, with only three demerits.[8]

The following year saw landscape and topographic drawing and English grammar added to the curriculum (after all, an officer or engineer had to be able to communicate graphically and verbally), and Stone again placed eighth academically while picking up only one demerit, which ranked him fourth among the entire corps of 223 cadets. At the annual final examination in June 1843, Stone came into contact for the first time with General Winfield Scott, America's foremost military hero, and a man who later would play a key role in his career. Standing well over six feet and resplendent in his major general's uniform, Scott cut an imposing and, to the cadets, frightening figure. Although he had not attended West

Point, one of his favorite activities was to chair the Board of Visitors. Despite the ordeal of having to recite before the imperious major general and other distinguished officers, Stone did well, again passing the exam. During his third year, the academics became more demanding, with instruction changing to chemistry and "natural and experimental philosophy," as the study of physics and the scientific method was called at the time, as well as more drawing. Cadet Stone did not do as well in these subjects, dropping to tenth in his class academically and accumulating five demerits. Cadets learned much about soldiering and service to their country during their fourth and final year at the academy. That year Stone excelled in ethics, which included constitutional and international law, rhetoric, and logic, as well as in artillery tactics, and mineralogy and geology, while performing just adequately in infantry tactics and civil and military engineering. He also took a nine-week course called the "Science of War." This subject, taught by America's leading military theoretician, engineering professor Dennis Hart Mahan, covered a wide variety of topics ranging from outpost duties to the organization, strategy, and tactics of armies. During the first two weeks of June, Stone passed his final senior examinations, and once again appeared before Major General Winfield Scott. In his last term he received six demerits. When the final order of general merit for the class of 1845 came out, he placed a very respectable seventh out of forty-one, just ahead of Fitz-John Porter. What thoughts must have been going through his head as he marched in the final dress parade and joined in the senior class celebrations. Like his fellow graduates, he anticipated putting behind him what George McClellan described as "four years of slavery"—the books, recitations, and drill, along with the stark barracks that had been his residence for that time—and going home before receiving a field assignment.[9]

The years he had spent at West Point not only provided Cadet Stone with a first-class military and engineering education but with opportunities for social, political, and spiritual development as well. He had cultivated personality traits that he retained throughout his life: honesty, forthrightness, loyalty, and tenacity. At the same time, however, he never lost a certain naiveté and lack of guile and intrigue. While the former traits would serve him well in future years, the latter would be the cause of many of his disappointments. Interactions with classmates of diverse backgrounds also motivated him to question his own standards and beliefs. By the time of his graduation, Stone had shed the Whiggish politics of New England and had developed democratic leanings like many of his fellow cadets from the Southern and Mid-Atlantic states. While at the academy, he also abandoned the Episcopalian traditions of his forebears and converted to Roman Catholicism, perhaps influenced by other Catholic cadets.[10]

Upon graduation, the cadets shed their gray uniforms for the blue and bullion of officers in the regular United States Army. Since the government allowed only a limited number of officers in each rank, many West Point graduates, including Stone, were given the status of brevet second lieutenants until permanent postings became available. Brevets were warrants authorizing the holding of a higher rank temporarily, but without the commensurate increase in pay. Post-graduation assignments depended on a cadet's academic standing. Those with the highest ranking in the order of general merit were given the opportunity to join the army's elite Engineer Corps, or the Corps of Topographical Engineers. Stone, however, finished just two levels below that needed for these prestigious appointments. Nevertheless, he still had options, unlike those unlucky cadets who ranked farther down in their classes and who would be consigned to the infantry or dragoons. Stone chose the Ordnance Department, a very small unit (at the time of the Mexican War, the department had only thirty-six officers) that tested, maintained, and issued weapons and supervised the production of arms and ammunition for the army. These tasks appealed to Stone's sense of order and organization. While usually a non-combat unit, the Ordnance Department staff could serve as artillerists during wartime, as the young West Point graduate would soon find out.[11]

Although he had placed near the top of his class, and had received an appointment on July 1, 1845, to the Ordnance Department, Brevet Second Lieutenant Charles P. Stone still could not report for service until a position opened up. Openings usually only occurred upon the death, resignation, promotion, or transfer of an incumbent. So, as the graduating class boarded the

Hudson River steamers to leave for their homes and await orders to new posts, Stone, along with a handful of other graduates, would have to return to West Point. During this hiatus, they were assigned to be acting assistant professors at the academy. Stone was technically an instructor in history, geography, and ethics, but in the fall of 1845 he taught only ethics to senior-year cadets. As the semester progressed, the ambitious young officer must have felt a strong sense of frustration when, one by one, his fellow temporary instructors received orders to report to their assigned units. Finally, Stone's turn came in December 1845, when orders arrived relieving him of his teaching duties. The following month he was granted leave, and he departed West Point. Early in 1846, Brevet Second Lieutenant Stone joined the Ordnance Department at its Watervliet Arsenal up the Hudson River just past Albany. He soon transferred to Fortress Monroe, along the southeast coast of Virginia, where he became the assistant to the commander of the fort's arsenal, Captain Benjamin Huger. During his time at Fortress Monroe, Stone learned as much as he could about ordnance work, including the handling of siege guns, all the while keeping abreast of the conflict between the United States and her southern neighbor, which was developing along the Texas–Mexico border, and awaiting the call to war.[12]

TO MEXICO
WITH
WINFIELD SCOTT

W HILE Stone was graduating from West Point and, as a newly minted brevet second lieutenant, lecturing cadets on the finer points of ethics, events were transpiring far to the west that would impact his career and that of many academy alumni. Since the presidency of Andrew Jackson, many Americans had coveted the Mexican province of Texas. The desire for Texas increased after the province gained its independence in 1836. Many Texans wanted annexation to the United States as well, but sectional issues regarding the expansion of slavery stalled the acquisition for almost a decade, during which Texas struggled as a republic. President John Tyler concluded a treaty of annexation with Texas in April 1844, but it failed ratification by the Senate. The following March, however, just prior to leaving office, Tyler signed a joint congressional resolution annexing the young republic. This act did not go unnoticed in Mexico, which for some time had let it be known that it would regard annexation of its former province as tantamount to a declaration of war.

In July 1845, to reinforce the American position, the new president, expansionist James K. Polk, dispatched General Zachary

Taylor and a small army into Texas as far as the south bank of the Nueces River. Since 1836, Texas had been engaged in a dispute with Mexico over its southern boundary. Texas claimed that it was the Rio Grande, while Mexico held that it was the Nueces. Now the United States was about to become embroiled in this controversy. Mexico's refusal to receive American diplomat James Slidell prompted President Polk, in January 1846, to order Taylor to advance to the Rio Grande, a move almost guaranteed to provoke Mexico into war.

The shooting along the Rio Grande started on April 25, 1846, and by the time word of the hostilities reached Washington, DC, and Congress declared war, battles were already being fought. At Fortress Monroe, Stone, while freed from the limitations of teaching, still found himself far from the action. As General Taylor and his force of regulars and volunteers fought its way into northern Mexico, he had to content himself with mastering the many details of ordnance work. Taylor's victories at Palo Alto, Resaca de la Palma, and Monterey led to the possibility that the war might be over before Stone had a chance to enter it. But, while Mexico had suffered some serious defeats, it was not yet out of the fight—not by a long shot. The country's charismatic leader, Antonio Lopez de Santa Anna, had betrayed the American offer to return him from exile if he would negotiate a peace. Once back on his native soil, Santa Anna took control of the army and set about organizing resistance to repel the invaders. Mexico's refusal to enter into a peace convinced Polk and his advisers that nothing short of an invasion of central Mexico and capture of its capital would be necessary to bring the enemy to terms. But who to lead this bold venture? While Zachary Taylor had won battles in northern Mexico, the president thought him incompetent and a future political threat. Other generals either lacked the knowledge and leadership capabilities, or were politically unacceptable. That left the army's senior general, Winfield Scott. Scott, who had served since the War of 1812, certainly possessed the ability to lead the invasion— indeed, he had already begun planning it—but he and the president disagreed on almost everything military and political, except for the necessity of invading central Mexico. Nevertheless, after weighing his options, and, despite his dislike for the general and considerable misgivings, Polk finally selected Scott.

The invasion plan for central Mexico that Winfield Scott had been preparing recognized that, in addition to infantry and cavalry, heavy artillery and siege guns would be required for success. To oversee the supply, management, and employment of ordnance for the expedition, he chose Stone's commanding officer at Fortress Monroe, Captain Benjamin Huger. Widely regarded as the army's leading authority on ordnance, Huger was a native of Charleston, South Carolina. He graduated from the United States Military Academy in 1825, and served in the artillery and as a topographical engineer before being given command of the arsenal at Fortress Monroe in 1832. That year he was transferred to the Ordnance Department with the rank of captain. Huger left Fortress Monroe in 1839 to serve on the army's Ordnance Board, and was a member of the American Military Commission to Europe from 1840 to 1841. He returned to command the Fortress Monroe arsenal in 1841, and was serving in that capacity when selected to be chief of ordnance for the Mexican invasion.[1] Scott's small Ordnance Department was responsible for inventorying and distributing weapons and ammunition, and for repairing field artillery and siege guns. Ordnance men might also staff batteries, if required. As a member of Huger's ordnance staff, Stone would be given the opportunity to play a key role in the upcoming campaign.

General Scott's conquest of Mexico began in the spring of 1847. Its success depended on the capture of the old colonial port of Vera Cruz on the Gulf of Mexico to serve as the base for the invasion. He knew he had to take the city and the adjacent fortress castle of San Juan Ulua and get his troops into the highlands before early summer when the yellow fever, known locally as "*El Vomito*," season set in along the swampy coast. An epidemic could decimate his army and leave the invasion defeated before it began. The need to capture Vera Cruz quickly prompted some of his officers to urge an infantry assault on the walled city and San Juan Ulua. Scott, however, reckoned that such an attack would result in an unacceptable number of casualties, a "butcher's bill," too early in the campaign. Instead he opted for a siege and bombardment, "the slow, scientific process," as he called it, to bring Vera Cruz to its knees. This strategy would require Huger's ordnance staff to bring a significant amount of guns, ammunition, and supplies from the country's arsenals by ship to the invasion site.[2]

Captain Huger assigned Second Lieutenant Josiah Gorgas the responsibility of transporting the siege train to Mexico. A West Point graduate, Gorgas had been with the Ordnance Department since 1841, and in late 1846 was serving as the assistant ordnance officer at the Watervliet Arsenal. The siege train consisted of weighty eighteen- and twenty-four-pounder cannon and howitzers, as well as ten-inch iron mortars capable of lobbing explosive shells in high arcs over enemy lines and walls. Along with the guns came implements, forges and bellows, stores, and one hundred thousand pounds of powder. In addition, an ordnance company of officers and enlisted men was to accompany the siege train. Following Huger's direction, Gorgas arrived in New York City in mid-December, and on the twentieth reported to the ordnance depot on Governor's Island. He sent orders to Stone to bring a detachment of enlisted men from Fortress Monroe to New York to join with his command in the formation of the company. By January 16, 1847, Stone and his men had arrived, and the siege train had been loaded on the transport *Tahmaroo*. A three-masted, square-rigged sailing vessel of 390 tons, the *Tahmaroo* had been launched at Blue Hill, Maine, in 1844 and employed in the trans-atlantic trade prior to her requisition by the government for transport service.[3]

Gorgas, Stone, and seventy soldiers also boarded late on the sixteenth, but the *Tahmaroo*, her passengers, and cargo did not depart immediately from New York. A lack of wind kept her at the quarantine station where she had been moored because of the powder on board. Finally, after three very cold and calm days, a breeze came up, and on the morning of January 19 the *Tahmaroo* was able to set sail and clear the harbor. By nightfall she was out of sight of land. With about six hundred tons of cargo on board, the ship rode quite low in the water, causing her to roll uncomfortably during the stormy weather that marked the first five days of the voyage. Seasickness plagued the majority of the enlisted men, who were quartered in the forward section of the vessel. Gorgas and Stone had small cabins in the slightly steadier stern. There they shared meals with the *Tahmaroo*'s master, Captain Sinclair, and the mates. As the ship sailed south, the storms abated, the sea calmed, and the troopers began venturing on deck for fresh air, sunshine, and some recreation, often in the form of singing. Gorgas and Stone had the run of the quarterdeck, where they often would lie down, enjoy the music and the ship's gentle rocking, and soak up the sun. By January 28, the *Tahmaroo* had reached the Bahamas, and, as she progressed through the warmer waters, was accompanied by schools of flying fish and flocks of sea birds. With seasickness now behind them, some of the soldiers and crew grew bored and mischievous. One of their favorite pranks was to steal food from the officers' and mates' mess. This pilfering resulted in the stationing of a guard in an unsuccessful attempt to protect the victuals. Exasperated, Gorgas and Stone sprinkled an emetic on the last serving of cheese before it, too, was stolen. This move resulted in a most unpleasant few days for the thief, or thieves, but it did end the larceny.[4]

Passing Cuba on February 1, the *Tahmaroo* entered the Gulf of Mexico. The weather warmed considerably and the sea smoothed even more. When not taking his *siesta* on the quarterdeck, Stone tried fishing, primarily for barracuda, which were abundant in the gulf. Gorgas and Stone also drilled their men, and directed them in inspecting and securing the cargo and cleaning the deck. To make the sailing more pleasant for all, they fumigated and doused with vinegar the crowded enlisted men's quarters every few days, and made the soldiers bathe regularly. At this point in the voyage, the members of the company had been cooped up aboard the *Tahmaroo* for a few weeks and were likely getting on one another's nerves, so the officers provided them with all they could eat, for, as Gorgas noted, "a full stomach makes a complacent disposition, while a hungry man is always ready for a 'row' "—or light-fingered with officers' food.[5]

The *Tahmaroo* made landfall along the northeast Mexican coast on February 8, and received orders to report to the Panuco River port of Tampico, which had already been occupied by American troops. Sailing at approximately nine knots, the ship reached the mouth of the Panuco River on February 17, and two days later, Scott arrived aboard the screw steamer USS *Massachusetts*. Among his entourage were Captain Huger and Huger's second-in-command, First Lieutenant Peter V. Hagner, a graduate of the West Point class of 1838 and career ordnance officer. Hagner had also been Huger's assistant at Fortress Monroe prior to Stone's arrival in 1846. He joined Gorgas and Stone aboard the *Tahmaroo*, which set sail the next evening for Lobos, a coral island about fifty miles south of Tampico, and the rendezvous point for the invasion fleet.

The following day, she anchored off Lobos and found some twenty-five ships already lying around the island, with more appearing almost hourly. The USS *Massachusetts* arrived later, and, in an attempt to change her position, narrowly avoided ramming the *Tahmaroo*. Twice the size of the *Tahmaroo*, had the *Massachusetts* collided with the heavily loaded transport, the latter ship would have sunk like a rock, probably taking most of Scott's Ordnance Department with her![6]

On February 24, the *Tahmaroo* set sail once again, this time in company with another ordnance transport: the *St. Cloud*. The ships were bound for the anchorage at Anton Lizardo, about twelve miles south of Vera Cruz. Their course took them just out of range of the guns at San Juan Ulua, where a sudden and furious storm— a *norte*, or "norther"—caused the vessels to heave to for a day. Finally, on the twenty-seventh, the *Tahmaroo* and *St. Cloud* anchored at Anton Lizardo. Among the first American ships to arrive, they waited for several days for the rest of the invasion fleet to gather. Unbeknown to Stone at the time, back at the War Department his promotion to regular second lieutenant became official on March 3, and he would no longer be just a brevet officer. By March 8, all seemed in readiness for the landing. All, that is, except the weather. Another severe norther delayed the invasion for another twenty-four hours. The next day was calm and clear, and crews and soldiers aboard the fleet at Anton Lizardo sprang into action. Benjamin Huger remarked in his diary, "Everything in motion—boats coming out, troops being transferred, steamers smoking . . ." Stone and the other ordnance men watched the spectacle, which continued well into the night, from the decks of their ships. What they saw was truly impressive, as one observer noted:

> The tall ships of war sailing leisurely along under their top-sails, their decks thronged in every part with dense masses of troops, whose bright muskets and bayonets were flashing in the sunbeams; the gingling [sic] of spurs and sabers; the bands of music playing; the hum of the multitude rising up like the murmur of the distant ocean.[7]

Scott's plan was for the infantry and mounted troops to secure the beachhead (which they did, surprisingly, with almost no resistance) and begin digging a trench line around the city, and for the engineers to locate the sites for the batteries prior to landing the siege train. The encirclement (or "investment") of Vera Cruz was known officially as Camp Washington and was completed on March 13. In preparation for this, on March 11, the *Tahmaroo* was directed to be brought up. As she set sail, Gorgas and Stone ordered the siege train out of the hold and made ready to go ashore. Before the vessel could get into position, however, the wind died and she was left wallowing in the swells. While Stone and the soldiers tried desperately to keep the ordnance stores already on deck from sliding about, Gorgas took a small boat to get help from the fleet. He found the steamer *Alabama* already getting under way to assist the *Tahmaroo*. Unfortunately, as she approached, the *Alabama* mistook another vessel for the becalmed transport and sailed right by. Once the *Alabama*'s captain realized his error, she had to come back, and it would not be until almost midnight that the *Tahmaroo* finally anchored off Sacrificios. But still the siege train remained on board, as the next day a surprise norther again stirred up the sea and prevented the landing.[8]

The morning of March 13, however, dawned calm, and the transfer of cannon and ordnance stores to the beach commenced. Later in the day, Gorgas and a detachment of soldiers went ashore while Stone remained aboard ship and saw to the safe loading of the guns and supplies onto the surfboats. Despite Scott's impatience to get his siege guns in position, the process went slowly, with as many as twenty surfboats landing at one time, each discharging a variety of commissary, quartermaster, and engineering property, as well as ordnance. Gorgas recalled the rather unorganized scene:

> Confusion . . . Quartermasters and wagon masters rode wildly along the beach, directing, urging, scolding and swearing. Mule teams of the most refractory kind, jammed round among each other and our piles of property. Volunteers lounged around in squads regaling passers with provincialisms. Here a huge sling cart with its pendant gun blocked the way—there a train of mules just landed wound about bewildered, to their destination.[9]

Despite the chaos, by the end of the day six pieces of heavy artillery had been landed and installed in a battery protected by sandbags and breastworks. It was hard work, with soldiers and some sailors wielding picks and shovels to construct more batteries and underground magazines, and hauling the heavy guns through thick chaparral and over high sand ridges. The labor continued for the next several days. Back on the beach, for volunteer soldier John Jacob Oswandel the uproar seemed to have subsided:

> I took a walk to the beach and saw one fellow still busy in land-ing cannons and mortars. Some are hauling cannon balls while others are hauling ordnance stores and provisions for the army and to different batteries.[10]

The busy fellow on the beach may have been Stone. By this time, he had bade farewell to the *Tahmaroo* and was assisting in moving and installing the mortar batteries while sporadic rockets and cannon shot from Vera Cruz and San Juan Ulua flew about. Although he did not receive a scratch, the young second lieutenant knew now that he was at war.[11]

On March 22, Stone was serving as officer of the day for the ord-nance company, seeing to the unit's daily routine, filling out morn-ing reports, and posting sentries to protect the stores. He may have been startled when, about four o'clock in the afternoon, cannon and mortar fire erupted from the American batteries. The reduc-tion of Vera Cruz had begun. The next day found Stone in the trenches, making sure that the fuses on the exploding shells were set to detonate at just the right moment and trying to keep the bat-teries supplied with powder and ammunition. In his efforts, Stone was hampered by another norther, which kept a sufficient amount of shot and shell and more mortars from being landed, and Scott lamented that his guns fired "languidly" because the batteries lacked the means to deliver a sustained pounding to the city, and he did not have the firepower to force San Juan Ulua into submis-sion. Once the wind died down and the sea calmed, however, the ordnance soldiers and sailors from the transports worked feverish-ly to unload and distribute the ammunition, and to get more mor-tars to the siege line. Their labors paid off, and the bombardment gained intensity.[12]

For the next forty-eight hours, American shot and shells from the trenches and from warships off the coast roared into Vera Cruz, destroying military works and civilian buildings as well. Scores of soldiers and noncombatants were killed and wounded. The fresh supplies and new mortars added to the devastation the Americans wreaked upon the city. Navy lieutenant Raphael Semmes, who would later become one of the Confederacy's leading naval officers and most daring sea raiders, had been assigned to the invasion force and was taken aback by the ferocity of the attack:

> Those horrid mortars of ours were in "awful activity." The demons incarnate, all begrimed with powder and smoke, who served them at this midnight hour, having received a fresh supply of shells and ammunition, since the lull of the norther, seemed to redouble their energies. . . . They gave the doomed city no respite.[13]

Edward Mansfield, General Scott's biographer, painted an even more frightening picture of the bombardment:

> Terrible was the scene! The darkness of night was illuminated with blazing shells circling through the air. The roar of artillery and the heavy fall of descending shot were heard through the streets of the besieged city. The roofs of buildings were on fire. The domes of churches reverberated with fearful explosions. The sea was reddened with the broadsides of ships. The castle of San Juan [Ulua] returned, from its heavy batteries, the fire, the light, the smoke, the noise of battle. Such was the sublime and awfully terrible scene, as beheld from the trenches of the army.[14]

The barrage continued as night fell, prompting one American officer to describe the appearance of the mortar fire and its possible consequences:

> Our mortars . . . poured in a perfect stream of shells into all parts of the City, the very thought of which makes me now shudder. The shells were filled with several pounds of pow[d]er and at night might be seen by their burning fuzes making their passage from the mortars . . . through an immense arc, rising very high and then descending into the denoted City & probably falling . . . into some house through the roof would there burst with an awful explosion, destroying whole families of women and children. It is horrible to think of.[15]

For a Mexican observer in Vera Cruz, the impact of the shelling was all too immediate:

The shooting continues without a break . . . mortars, howitzers, cannon, batteries from the smaller ships . . . all targeting the city center. . . . Instantly the horrors of a fort under bombardment begin. . . . While a wounded man is being operated on, a shell explodes and plunges the room into darkness; when the lamps are relit, the patient is torn to pieces and many more are wounded or dead. Scenes of agony and blood follow one after the other, which we must decline to describe to avoid a litany of horrors.[16]

Eventually, the Mexican guns fell silent. Rumors spread in the city that Santa Anna was on his way with a relief column, but when they proved to be just that—rumors—the military and civilian authorities in Vera Cruz faced a desperate situation. They searched for an honorable end to the destruction, but Scott let it be known that he would accept only one solution: surrender. Consequently, on March 26, the white flag appeared above Vera Cruz and the bombardment stopped. By terms of the surrender, the Mexican soldiers stacked their arms and marched out. The Americans occupied both the city and San Juan Ulua on March 29. In all, the Americans expended some three thousand ten-inch mortar shells, two hundred howitzer shells, and 3,500 round shot in the capture of the city. Casualties amounted to sixty-seven, including two officers killed. Mexican losses numbered in the hundreds, most of them unlucky civilians.[17]

It is impossible to fathom just what Stone thought as he wandered the streets and plazas of the battered city. While he was a professional soldier, he also was known to have a sensitive nature. As with many of his fellow officers who previously had not really witnessed combat, he likely experienced a sense of pride that his army had invaded a foreign country by sea and gained such a signal victory in so short a time with so few casualties on its part, tempered with a helping of guilt and grief over the suffering inflicted on the general population. No doubt, however, for Stone and his junior officer colleagues, the ruined buildings and the dead and maimed civilians they encountered caused the war to become all too real, all too quickly.

The work of the ordnance company did not diminish with the fall of Vera Cruz. Ordnance now had the responsibility of shifting the artillery from the soft sands of the trenches to firmer ground for easier transportation in preparation for the army's move inland.

Stone and his men also had to deal with the weaponry and ammunition captured from the Mexicans. About four hundred cannon, along with shot, shell, and powder had fallen into the hands of the Americans. The soldiers would spike or otherwise render unusable most of these guns. Some would be sent back to the United States as "trophies," while still others would be made serviceable and transferred to Scott's army, which had not yet received its full complement of artillery. In addition to the big guns, thousands of muskets and other hand weapons were taken at Vera Cruz.[18] Lieutenant Hagner described the cache of weapons the Mexicans left behind:

> We get plenty of trophies, you see many of these fine old brass pieces from Spain—and pieces of ordnance—old Bombards by name—which must have been made as early as 1620. We get too, many fine pieces of our own muskets—made at West Point Foundry for several years past with thousands of shot, shells & . . . English guns, equally well supplied and of the latest patterns.[19]

While few of the antiquated firearms met the standards of the American army, they all had to be inventoried and safely stored, a tedious and time-consuming task.

In early April, General Scott ordered his forces to get ready for their advance westward from Vera Cruz to Mexico City. The *Vomito* season was fast approaching, and Scott wanted to move as far into the highlands as possible to escape the disease, and to position himself well inland before Santa Anna could assemble enough troops to try and push the Yankee invaders back to the sea. On the eighth, just ten days after occupying Vera Cruz, the first units of the American army marched out of the city following the "National Road" toward the Mexican capital. Ordnance, however, still had inventory and repairs to do, and was having difficulty procuring enough draft animals to haul the cannon, mortars, and store wagons, since few had been brought ashore with the invasion. Finally, by April 18, the work had been completed and animals rounded up, and the siege train with Huger, Hagner, Stone, and the ordnance company was on the road west to join up with the rest of the army. Their route took them into the highlands and provided a welcome change of scenery and climate from the swampy, mosquito-infested Gulf Coast. Raphael Semmes recalled:

It was delightful to inhale the morning air, as it came to us, lowered several degrees in temperature by the preceding night, and charged with the dewy perfume of flower and shrub. . . . Cacti abounded in many beautiful and novel varieties . . . We occasionally passed magnificent shade trees on the roadside, inviting the sun-burned traveler to pause and rest his weary steed.[20]

But the army had little time to rest its horses or its feet. So far, it had almost been too easy. The Mexicans at Vera Cruz had surrendered practically without a fight, and now haste was the order of the day. Haste to defeat Santa Anna. Haste to the gates of Mexico City.

The march inland, however, was not without its dangers. Mexican irregulars and guerillas sniped at the invading force, and it would not be long before the Americans met organized resistance. At Cerro Gordo, about fifty-five miles from Vera Cruz in the hills above the Plan del Rio, Santa Anna dug in over five thousand foot soldiers, cavalry, and artillery to stop the Yankees. On April 18, the same day that the siege train left Vera Cruz, the Americans routed the Mexicans in a bloody battle. While the ordnance company did not have the opportunity to participate in this fight, as in the case of the Vera Cruz aftermath, it was responsible for the inventory and disposition of weapons left behind by the enemy. Since the Mexicans fled the field of battle in quite a hurry, they had no time to tend to their dead, leaving a horrifying scene for Gorgas and Stone as they collected the discarded ordnance. At Vera Cruz they had witnessed the effect of artillery, but here at Cerro Gordo they saw the detritus of hand-to-hand combat.[21] On a more pleasant note, Stone had the chance to leave the National Road and visit one of Santa Anna's haciendas (probably Lencero, about eight miles from Cerro Gordo). Along with ten cavalrymen, he wandered the grounds and rooms of the estate, making sure that the party did not damage any of the fine furnishings or palatial mirrors. The expedition did encounter a few Mexican lancers, but in the ensuing skirmish neither side suffered any casualties and both went their separate ways.[22]

In the meantime, the army continued its rapid advance westward, taking the towns of Jalapa and Perote, and eventually occupying Mexico's second-largest city, Puebla. With each of these victories, however, more and more enemy weapons fell into the hands of the Americans, further compounding the work of the ordnance company. It also faced the tasks of repairing damaged arms and keeping the army stocked with munitions over an ever-extending supply line. Despite the long days and hard work, Stone and his fellow officers enjoyed a fairly pleasant mess. When possible, they slept in houses or other structures, rather than in tents. A dinner cooked in camp might consist of a first course of vermicelli soup and rice, followed by a main dish of greens and a serving of meat, with a concluding course of pear, melon, sugar, and sherry. In addition, they were able to sample some of the local delicacies, such as tamales, which artillery captain Robert Anderson pronounced to be "*muy bueno.*" [23]

Prior to the occupation of Puebla, Stone again saw action in a skirmish at the village of Amozoque, about twelve miles east of the city. He and the siege train were assigned to General William Worth's division, which arrived at Amozoque on May 13. There he halted to allow a supply train and small brigade under General John Quitman to catch up, and to give his men the chance to rest, bathe, brush the trail dust from their uniforms, and clean their weapons. A veteran of the War of 1812 and the Seminole Wars, and former commandant of West Point, Worth appreciated a soldier's good appearance and wanted to make a presentable entrance into Puebla the next day. On the morning of the fourteenth, however, a drummer boy who had wandered beyond the American picket line returned with news that he had seen a sizeable Mexican cavalry force. Indeed, there were between 2,500 and 3,000 lancers under Santa Anna lurking nearby. After Cerro Gordo, he had sent his surviving infantry and artillery to Mexico City, but had kept the lancers, mounted on horses roughly requisitioned from Puebla's civilian population, on the road to make whatever mischief they could. In this case, Santa Anna sought to ambush the supply train and hinder Worth's advance. He hoped also that a victory might motivate the jaded citizens of Puebla to ignore his recent theft of horseflesh and rally to his side against the Americans. Once the warning was given, however, Worth's troops moved quickly to foil the attack. One writer described their response to Santa Anna's sudden appearance at Amozoque:

> Several of the junior officers [perhaps including Stone] visited
> the principal church in the little town. . . . While in the church,

much interested in what they were seeing, and recipients of kindly attention of a padre, they heard the long roll from every drum in the division, continuous and incessant, creating not merely surprise, but alarm; and instantly they were seen crossing the plaza in all directions as fast as they could run to join their commands.[24]

The Americans sent some infantry, cavalry, and artillery units to meet the Mexicans. After a few shots were fired, Santa Anna realized he had lost the element of surprise, and led his lancers away to Mexico City. Although it is unclear what Stone did at Amozoque, his record was credited with participation in the engagement.[25]

The day after the skirmish, Worth marched unopposed into Puebla. Despite their attempts to clean up, the American soldiers did not impress the city's curious residents, who turned out to witness the occupation. They were accustomed to the varied and colorful uniforms and shakos of the Mexican infantry regiments and lancers, the shiny helmets and breastplates of the cuirassiers, and the tall bearskin hats of the Grenadier Guards of the Supreme Powers. The simple gray and blue field attire of the Americans, with their low-billed caps, seemed drab in comparison. One journalist described the invaders as "decrepit in appearance, dressed in uniforms poorly made and ill fitted," while another observer saw them as "coarse and clownish men." Their actions also seemed rather unmilitary, with most stacking their weapons in the plaza and then going off unarmed to find food and water while the remainder lay down and took a nap.[26]

With the arrival of Scott and the rest of the army at Puebla a short time later, the campaign went into a hiatus for a couple of months. Senior state department clerk Nicholas Trist had arrived in Mexico and was trying to negotiate a peace with the Mexican government, while Scott awaited replacements for many of the volunteers whose one-year enlistments had expired and who had chosen to go home. The American regular troops and remaining volunteers, in the meantime, rested and explored the plentiful natural and cultural sights in and around Puebla. A number of the officers found the region to be one of the most attractive in Mexico. Just about a day after Worth's entrance into the city, however, Stone became very ill with a fever and saw little other than the inside of his quarters. This was the first time he had been sick since leaving New York. The malady lasted almost through the army's summer stay in Puebla, although he did venture out to attend mass with other officers on July 4.[27]

By early August, the Americans had been reinforced and resupplied, and it had become clear that neither negotiations nor a direct bribe to Santa Anna himself would bring a stop to the fighting. On August 7, therefore, Scott resumed his offensive. In order to bolster his forces for the attack on Mexico City, he ordered the abandonment of Jalapa, Perote, and other outposts, with their garrisons, ordnance, and stores rejoining the main army. This left Puebla as the only occupied town on the road from Vera Cruz. The American sick and wounded would remain there, along with a small force to guard the city and care for them. While not quite as dramatic as Cortez burning his ships before advancing on Tenochtitlan in 1519, by abandoning his supply and communication routes with the Gulf Coast, Scott made it clear to his troops and to the Mexicans that there would be no turning back—no substitute for victory.

In the meantime, Stone had recovered sufficiently to accompany the army as it marched out of Puebla. He was still with the siege train, which had been attached to Worth's division. When the troops reached a vantage point above the Valley of Mexico, they were presented with a spectacular vista. Second Lieutenant Simon Bolivar Buckner, with the Sixth Infantry Regiment, described the scene that greeted the invaders:

> The Valley of Mexico, shut out from the surrounding world by a range of lofty mountains which completely encircles it, was spread out far below us. . . . From the height at which we viewed it, we gazed upon the many beauties which it unfolded; upon the highly cultivated fields, upon the villages and cities, and spires, upon the distant mountains which bounded the horizon, and which were mirrored by the smooth surface of the lakes which reposed tranquilly beneath.[28]

Artillery captain Robert Anderson also wrote of a seemingly idyllic march down into the valley:

> Every turn of the road now opened to us a new or more extensive view in which the pictures were formed. Every variety of green that could be formed by the varied light and shade of

passing clouds and by real difference of shade, with mountains here, nearly in the foreground, there in the distance, and beyond, limiting the view; and lake, in this part almost un-distinguishable from the grass and slime which nearly covered it, to the clear water, in which the shadows of passing clouds were visible, the pictures studded with haciendas, some traced out by their huge mud walls enclosing immense courtyards, like fortifications, villages with churches, etc., presented views which were charming to those who hoped that there lay the City from which they must return to their beloved homes. . . . Indeed, independent of everything, the scenery was beautiful. The descent was rapid, and the view of our Division, which could be seen distinctly by looking ahead and in the rear . . . as it wound its way down compactly and rapidly, was the most beautiful panorama picture I ever saw; our wagon train extended two or three miles and could be seen with their white tops passing through the trees which shaded the road.[29]

The view may have been superb, but Stone had little opportunity to enjoy it. He spent most of his time trying to keep the bulky wagons with their obstreperous draft animals, heavy guns, portable forges, shot, and shells on the road and moving at pace with the rest of the army. This was not always easy, as another artillery officer noted in his diary:

That road was most horrible & as the train was in front of us, we were much delayed by it & on an average did not advance more than three hundred yards at a time without a halt. The Mexicans had obstructed the road in several places by rolling down immense rocks from the hills & it required much labor to remove these obstacles.[30]

Occasionally, though, Stone did venture away from the siege train. On August 11, Huger spied a Mexican lancer looking "impudently" at the Americans and sent Stone and a few soldiers in a futile attempt to apprehend him. He had better luck a few days later while accompanying one of the parties reconnoitering the best route to Mexico City. Stone was able to take prisoner a Mexican officer whom he brought proudly to the division headquarters. By this time the army was on the valley floor, and the beautiful vistas had changed to scenes of poverty and squalor. The soldiers also began hearing disturbing reports that, while they were stalled in Puebla, Santa Anna had thrown up formidable defenses to oppose them. Rather than remaining on the main highway along the northern shore of Lake Chalco and facing the Mexicans' strongest fortifications at El Penon, therefore, Scott chose the less-used southern route around the bottom of Lakes Chalco and Xochimilco. The Americans slogged along the muddy lakeshores, and in a series of engagements dislodged the enemy from its defensive positions. Stone was credited with participation in one of these battles, at the agricultural village of Padierna (usually called Contreras) on August 20. Although the general did not mention Stone in his report, the lieutenant may have been involved in the moving of artillery and ammunition, along the path hacked through the almost impenetrable lava bed known as the Pedregal, by Captain Robert E. Lee of the engineers and five hundred troops. The move through the Pedregal allowed the Americans to surprise the Mexicans and shell their flank, contributing significantly to the victory.[31]

As the smoke of battle cleared, Stone and members of the ordnance company were assigned to locate the twenty-two cannon said to have been left behind by Mexicans retreating from Padierna, as well as additional field pieces that had fallen into American hands during their victory at Churubusco, also on August 20. One of these guns had "Death to the Yankees" emblazoned on its barrel, while another had "Kill Yankees" painted on it—indications that the Mexicans still had some fight left in them. Two days later, Huger received an order to assemble all the enemy weaponry captured since Puebla. This assignment kept Stone busy for the next couple of weeks. With twenty wagons, he went over the Padierna battleground and other locations collecting guns and ammunition. But it was not all drudgery. On August 25, much to Stone's delight, his companion from their West Point days, Second Lieutenant Fitz-John Porter, joined the ordnance officers' mess, and they enjoyed each other's company for several weeks.[32]

During the time Stone and the ordnance company were dealing with captured cannon, Scott had entered into an armistice with the Mexicans hoping at best to end the conflict altogether, or at least to allow his army the opportunity to resupply. This hiatus in the fighting also allowed the ordnance company to complete the retrieval, inventory, and relocation of all thirty-seven enemy guns along with implements and ammunition. By September 6, however- er, it was clear that Santa Anna was in no mood for peace, and preparations began for the final stage of the campaign. Scott's sen- ior officers offered opinions on the best way to capture Mexico City, and the general listened to them all before revealing his own plans. These were based on the latest intelligence (inaccurate though it may have been), and involved a feint along the southern approaches to the city while the main force moved north from the army's headquarters at Tacubaya, located southwest of Mexico City, and the last town before the capital. Worth's division would take the old grist mill known as "*El Molino del Rey*" and the nearby stone building referred to as the "*Casa Mata.*" In this complex, the Mexicans were said to be melting down church bells and casting them into cannon. Next Scott intended to seize the hilltop fortress of Chapultapec Castle, home of Mexico's military academy, and from there enter the city through one of its western *garitas* or prin- cipal gates with strong blockhouses.[33]

The Battle of Molino del Rey began early on the morning of September 8. Huger had command of two twenty-four-pound siege guns, which he placed on a ridge between Tacubaya and the old mill, about six hundred yards in front of the Mexican lines. Here, he hoped to "shake" the enemy in support of the infantry assault. Just prior to six A.M., Huger's battery began shelling. An observer described what followed:

> Very soon after the dawn of the day, the report of Huger's guns opening on Molino del Rey, gave the signal for attack. So heavy were the discharges, that in a short time masses of masonry fell with tremendous noise and the whole line of intrenchments [sic] began to shake.[34]

The Mexicans returned the artillery fire but did no damage. After only a few rounds from the American guns, the enemy was considered to be sufficiently "shaken," and Huger stopped his bombardment as the foot soldiers began moving forward. When the fighting started, Stone was unloading wagons of captured ammunition at Mixcoac, a village south of Tacubaya, where General Gideon Pillow's division was encamped. At the first sound of the artillery fire, he hurried to the battlefield and was given charge of one of the twenty-four-pounders. Stone moved the gun to the southern end of the mill, where he beat back a surprise attack from Chapultapec on an artillery battery commanded by Captain Simon Drum. He also silenced a Mexican field piece that had been hindering the American assault. As Huger put it, "a few shot from Mr. S's 24 [pounder] caused them to retire, or 'bamos.'"[35]

After two hours of bloody hand-to-hand combat, Scott's troops had driven the Mexicans out of the mill and Casa Mata. Much to their chagrin, however, the Americans found no evidence that their opponents had been manufacturing cannon in either building. Nonetheless, they considered the fight to be a victory, and this time, Stone's efforts did not go unnoticed. Because of his "gallant" actions at Molino del Rey (and perhaps at Padierna/Contreras as well), Scott appointed him a brevet first lieutenant. Despite the honor, his duties remained the same. On the day after the battle, Scott ordered Huger to make Mixcoac the army's general depot for arms and ammunition in preparation for the assault on Mexico City. So, it was back to Mixcoac for Stone to assemble and organize the American ordnance, as well as the captured Mexican artillery, which Scott estimated had tripled the number of siege guns at his disposal.[36]

Before they could move on Mexico City, the Americans had to clear the fortress of Chapultapec Castle of its defenders. If Chapultapec remained in enemy hands, Scott reckoned that his planned assault on the city's west gates would require an awkward and dangerous roundabout movement exposing his flank to attack. He hoped also that the loss of Chapultapec might finally convince the Mexicans to surrender, removing the necessity of an attack on the heavily defended *garitas* and possible house-to-house fighting in the city. Huger suggested to Scott that the fort might be reduced by a heavy bombardment alone. This idea appealed to the general. The Mexicans' stand at Molino del Rey demonstrated that they could not be expected to turn tail and run at the first American volley, and charging the castle's stout walls could result in another day of carnage. The first phase of the attack, therefore, would involve an attempt to pound Chapultapec into submission. Huger began preparing his heavy guns for the barrage. Along with Robert E. Lee, he superintended the placement of the batteries, but the work did not go without incident. A mortar and a twenty-four-pounder overturned while being moved into position. Stone was given the assignment of righting and mounting the toppled pieces, tasks that took him well into the evening of September 11. In addition, he was placed in command of battery number four, the last of Huger's batteries that would shell Chapultepec the next morning. Comprised of a single ten-inch mortar, battery four was at the Molino del Rey, about 1,200 yards from the castle's walls.[37]

Acting on Scott's instructions, Huger ordered the batteries to open fire early on the morning of September 12. Batteries one and two began the shelling, with three and four soon joining in. The Mexican commander, General Nicholas Bravo, reported that, at first, the Yankees' aim was bad, although later their guns did much damage to the castle and caused many casualties. Stone's battery proved particularly damaging, inflicting severe injuries to Chapultepec's roof. By noon the Americans had silenced the enemy's cannon, but despite barrage after barrage, the tricolor flag still flew over Chapultepec as the sun went down. Nightfall brought the pounding to a halt. The American assessment of the barrage's effectiveness differed markedly from that of General Bravo. Inspector General Ethan Allen Hitchcock stated that, despite fourteen hours of almost constant bombardment, "no impression seems to have been made upon the main defences," and Major J. L. Smith, chief of the American engineers, and Robert E. Lee, Scott's most trusted engineer, both began having doubts as to the ability of artillery alone to destroy the Mexican position. After a review of the day's work, Scott decided that only an infantry assault would accomplish his goal. At 5:30 A.M. on the thirteenth, the American siege guns roared again, although this time the firing lasted only two and a half hours, until the infantry began its advance. Stone and the ordnance men and artillerists watched as the blue-and-gray–clad soldiers ascended the hill and scrambled over the parapets. An hour and a half later, the battle was almost over and the American flag waved above the castle. Everywhere the Mexican defenders could be seen dead, injured, captured, or trying to make their way back to Mexico City. In summarizing the morning's action, Scott called out the "decisive" role of the siege batteries, and later commended Stone personally for his "highly effective and distinguished" work and promoted him to brevet captain.[38]

The morning's victory, however, did not end the fighting. From Chapultapec, the Americans surged on to the western *garitas* of San Cosme and Belen. Hagner and Stone accompanied Worth's assault on San Cosme, and, as the infantry attacked the barricaded gate, they manned a gun which lobbed shells into the enemy's capital. By the end of the day, both *garitas* had fallen and the Americans were on the verge of occupying Mexico City. As the exhausted soldiers bedded down, Worth ordered Huger to give the enemy a "good night" message. In compliance with that order, he brought up to the edge of the Alameda Park a twenty-four-pounder and a ten-inch mortar (likely commanded by Stone) and fired a few shells that landed in the central plaza near the National Palace. The next morning found the city empty of hostile troops. Worth's message had had its effect, for overnight Santa Anna, after emptying the jails, had taken what remained of his army and evacuated north toward the town of Guadalupe Hidalgo. Civil officials surrendered the city. So, instead of days of desperate block-to-block and house-to-house combat, Scott and his army prepared for a triumphal entry.[39]

Although he had no way of knowing how the conquerors would be received, an elated General Scott donned his conspicuous dress uniform for a formal entrance into Mexico City at eight A.M. on September 14. Brevet Captain Stone and the ordnance company took a prominent place in this parade to the central plaza as part of the siege train that marched at the head of Worth's division. The event went off without a serious disruption, although for days thereafter, disgruntled Mexicans and freed convicts sniped or threw paving stones at American soldiers from houses and rooftops. Eventually, the city quieted down, save for the occasional beating or stabbing of drunken and abusive troopers, who usually hailed from the volunteer regiments. The ordnance staff quickly settled into a routine of arsenal work—repairing damaged weapons and inventorying and restoring captured guns. Most of this work was carried on in the dark and dirty old fort known as the "*Ciudadela*," or Citadel, near the Belen *garita*. In spite of their tiring and grimy toil, Huger kept the Ordnance Company well drilled and of "a very respectable appearance." He boasted often that the unit was one of the best-dressed companies in the invasion army. Stone and his brother officers found comfortable accommodations in a mansion belonging to one of the city's wealthier men. In their spare time, they enjoyed playing billiards, rummy, whist, and horseshoes. Some engaged in amateur theatricals while others went to the theater (although few could understand much of what was being said in Spanish), dance halls, and bullfights. As officers, they also attended a number of military social events, usually formal dinners with officers of other regiments. Stone hosted one such affair in his quarters on September 30 in celebration of his twenty-third birthday.[40]

While the army's senior officers occupied palatial residences that allowed them to relax and host splendid social events, their juniors initially had no such location for relaxation or entertaining outside of their lodgings. That changed in October when a group of seven officers, including Stone, organized the Aztec Club to establish a "resort" for themselves and their colleagues. Basically, it was to be an officers' club where members could enjoy leisure time together and purchase good food, drink, and cigars at reduced prices. The initiation fee was twenty dollars, not an inconsequential amount at a time when a second lieutenant of ordnance (Stone's actual rank) received a salary of only thirty-three dollars a month. The club's first president was General John Quitman, whom Scott had appointed as the military governor of Mexico City. Membership grew and quickly came to include Stone's West Point classmates Fitz-John Porter, Henry Coppee, and Barnard Bee, along with fellow ordnance officers Huger and Hagner, and men who would gain fame in the country's next major war—among them: Ulysses S. Grant, Robert E. Lee, Pierre G. T. Beauregard, George B. McClellan, John Bankhead Magruder, and James Longstreet. The Aztec Club secured quarters in the former town home of Jose Maria Bocanegra, who had been interim president of Mexico as well as interior and foreign minister and treasury minister. This eighteenth-century mansion was located just off the central plaza and behind the great cathedral, and had been built for the viceroy of New Spain. But now its stately halls echoed with the heavy boots and spirited conversations of the foreign conquerors.[41]

During the occupation of Mexico City, American soldiers were cautioned against wandering about alone, or even in small groups, lest they fall victim to an assault or worse. This warning, however, did not stop Stone and his colleagues from exploring the Valley of

Mexico and its environs. The great cathedral on the plaza must have awed the Catholic Stone, who was described by a fellow officer as being a pious man, as may have the artwork in the city's palaces and the relics of the Aztecs. But one location so fascinated him that he took bold action. At just over 17,800 feet, Mount Popocatepetl loomed over both Mexico City and Puebla. An active volcano, it was covered in snow year round, and, in the 1840s, still had glaciers on its flanks. It occurred to Stone that the conquest of Mexico would not be complete until the American flag was planted on the summit of this important landmark of the defeated nation. Consequently, in the spring of 1848, he organized an expedition to climb the peak.[42]

The ascending party consisted of twenty officers from various regiments, some enlisted men, a few civilians, and Mexican packers. The expedition rendezvoused at the Citadel and left Mexico City with an escort of dragoons on April 3. Traveling southwest some forty-three miles, passing through Mira Flores and Amecameca, the party arrived at the village of Ozumba at the foot of Popocatepetl. From there, the climbers and packers proceeded on horseback up the mountain to what was called the *vaqueria*, a corral and line shack for cowboys working cattle in the high pastures. They spent the night there, pelted by rain and later snow, which came through the shack's largely absent roof. Although they had not gotten much sleep, in the morning the climbers were eager to trek to the summit. Up the expedition went on foot, passing the vegetation line until snow clouds and violent winds blocked further progress and the soldiers retreated back to the *vaqueria* to try again the next day. Some climbers nursed frostbitten fingers and one officer suffered a frozen nose. Before dawn, however, everyone was afflicted with swollen and burning eyes, the result of the previous day's glare and wind. Only bathing their sore orbs in ice water provided any relief. Another ascent seemed impossible. Gradually sight returned to some of the party, and they led those whose eyes were still swollen shut and bandaged down to Amecameca, where they spent the night. By the following morning, all had recovered, but about half, including Grant, decided that they had had enough of the mountain adventure and went sightseeing elsewhere. Stone and the remainder picked up some climbing staffs and dark glasses and retraced their steps to the *vaqueria*. At about 2:30 A.M. they left on a final attempt to climb the volcano. Although it was calm and clear, the climbers battled deep snow along with the cold thin air on their way up Popocatepetl's slope. Some of the party had trouble concentrating, while others could take only a few steps before having to stop and rest. Finally, with blue lips and a pounding headache resulting from altitude sickness, Stone bounded to the summit, planting the Stars and Stripes there at a little after ten in the morning. He was followed by five other officers, three enlisted men, and an English civilian. They remained for about forty-five minutes, exploring the rim, picking up lava samples, and becoming nauseated by the sulfurous plumes of smoke the volcano emitted. The view, however, was astounding. They could see clearly both Mexico City and Puebla, as well as the countryside in between. On their way back to the *vaqueria*, his fellow officers had to prod an exhausted Stone not to lie down in the snow but to keep going. That night, the climbers enjoyed a well-deserved sleep by roaring campfires. They arrived back in Mexico City by April 14. On their way down the mountain, local Indians had refused to believe that they had made it to the summit, since only devils could do that, but these men grew no horns. Once in Mexico City, colleagues expressed sympathy for their failure until presented with rocks and other evidence to the contrary. Then it was heartiest congratulations all around.[43]

The conquest of Mount Popocatepetl marked the high point, both literally and figuratively, of Stone's time in occupied Mexico City. Like most American soldiers, he had tired of the routine of occupation and looked forward to returning home. A treaty between the United States and Mexico ending the war had been drafted in the town of Guadalupe Hidalgo and signed in early February 1848 and ratified by both countries by May 19 of that year. It called for the American army to leave as soon as possible following ratification. The evacuation of Mexico City began in late May and ended on June 12, with the troops mostly retracing the route along which they had fought their way a year ago. Huger had departed some time earlier, so it was up to Stone and the other junior officers to bring back the siege train. Despite their high spirits, the march from Mexico City to the gulf was no walk in the park. The ordnance company still held the responsibility of transporting and keeping in good condition the American guns and captured artillery that had not been turned back over to the Mexicans. But there were fewer of them to do the work. Ten of the army's ordnance men had died in battle during the war, while another twenty passed away from accidents or illness. After a weary journey, Stone and the siege train arrived in Vera Cruz, where few soldiers wanted to linger due to the presence of the *Vomito*. Transports were leaving daily, but the train still had to be loaded aboard ship. Stone finally arrived back in New York in mid-summer of 1848. He had been away a year and a half. During that time, like most of his fellow junior officers who survived the conflict, Charles Pomeroy Stone had been transformed from an inexperienced and naive second lieutenant into a veteran soldier with an appreciation of his triumph, firsthand knowledge of the art of war, and distaste for its rigors and tragedies.[44]

CALIFORNIA GOLD

ONCE back in the United States from Mexico, Lieutenant Stone awaited his next assignment. While many of his colleagues in the dragoons, infantry, and even the artillery either went west to serve at posts on the frontier or were detailed to coastal fortifications, he knew that in all likelihood he would find himself back doing routine ordnance work. And that is just what happened. He must have been disappointed to receive orders to report to Watervliet Arsenal as assistant ordnance officer, the same position he had held at Fortress Monroe prior to the war. Stone had more in mind for his career than a junior staff position in a department that would probably stagnate in a peacetime army. Mexico had been a great adventure. He had experienced combat and proven himself under fire. Now he yearned for more exciting service than inventorying muskets, inspecting stores, and repairing cannon. Little wonder, therefore, that when the opportunity presented itself, he applied for a leave of absence in July 1848 to study foreign military systems. Such junkets were not unusual in the antebellum army, as there was an abundance of junior officers and a shortage of assignments for them. As a result, the War Department encouraged them to broaden their professional horizons by observing European armies and their tactics. He obtained his passport in September, and was soon at sea again, this time on a packet heading across the Atlantic.[1]

Stone's grand military tour came at a critical time. The liberal insurrections and nationalist conflicts of 1848 dragged on in western and central Europe as armies of revolution and reaction swept across the continent. After spending a month in Paris studying the French military establishment (still considered the finest in the world) and renewing his fluency in the French language, he was off to observe the latest methods and implements of war. New elements were being introduced that would alter forever the conduct of warfare—principally the use of railroads to transport large numbers of troops quickly across long distances, and of the telegraph to provide near-instantaneous communication between headquarters and the battlefield. Both of these would be key facets in America's Civil War a decade later. At the time of Stone's visit, industrialization was also making itself felt in both civilian and military affairs. Large private factories, operating under contracts from national governments, were replacing army arsenals for the development and production of most weapons. Mass-produced breech-loading rifles were beginning to be supplied to the major powers, and, as an ordnance officer, Stone would have been keen to study the German firm of Krupp, which had begun supplying the Prussian army with new steel (rather than iron or bronze) artillery pieces.[2]

The ways of conducting battles were also changing. While the Napoleonic model based on outflanking and enveloping an enemy by massed volleys and bayonet charges, and the use of cavalry as shock troops, still held sway, these tactics were coming into question. Improved, longer-range, and more accurate rifles and artillery, along with more efficient and deadly ammunition, made closely packed infantry columns and mounted soldiers prime targets, encouraging some generals to try to break an enemy's center rather than envelop them. The idea of the "breakthrough" would come to dominate tactical thinking in Europe and America through the First World War. Stone would witness these changes as he moved about the Continent. In visits to battlefields and the laboratories and factories of arms manufacturers he may have traveled with his Mexican War comrade, Lieutenant Peter Hagner, who was in Europe on an assignment to investigate ordnance systems.[3]

During his time in Prussia, Stone also gained an audience with the renowned geographer Alexander von Humboldt. As the two men talked, Stone shared his experience on Mount Popocatepetl. Impressed with the young officer's account, the aged explorer arranged for him to attend a dinner with Prussian king Frederick Wilhelm IV, where he related the story to the monarch.

Stone's almost two-year journey abroad took him well beyond the usual tour of western and central Europe. He went on to the Middle East, visiting the declining Ottoman Empire to observe its campaigns in Syria, and then to the empire's near-autonomous province of Egypt. He lingered in this ancient land, studying the development of its new standing army, equipped with modern European weapons and trained by westerners. This force differed from previous Egyptian armies in that it was based on conscripts drawn from throughout the province rather than from a motley assortment of mercenaries, Islamic militias, and royal household retainers. Most of the officer corps, however, remained Turkish and Circassian, an issue that would rankle Egyptian nationalists for the next two decades. It is unclear if Stone perceived this dissatisfaction during his visit, but certainly if he could not have imagined it at the time, the makeup and leadership of Egypt's military were challenges he would later have to face.

After having been abroad for some eighteen months, Stone's transatlantic adventure ended in 1850. He returned to the United States and once again found himself an assistant ordnance officer at Watervliet Arsenal. His prospects soon improved, however, when he was placed in command of the arsenal at Fortress Monroe. But Stone's tenure in Virginia did not last long. Late in the year, he received unexpected orders sending him to the other side of the continent.[4]

In going to war with Mexico, one of President Polk's motivations had been the acquisition of California. A revolt against the Mexican government by Anglo settlers, combined with a successful invasion by the American army, navy, and marines, secured the northern portion of the province (Alta California) for the United States in 1846. Initially, Alta California was administered by the army, even after its cession by Mexico as part of the Treaty of Guadalupe Hidalgo in 1848. The discovery of gold in that same year accelerated the territory's settlement, so that by 1850, it had sufficient population and wealth to join the Union as the thirty-first state. The army continued to play an important role in California after statehood: protecting it and the rest of the Pacific Coast from foreign threats; subduing Native Americans who resisted displacement from their traditional lands by miners, farmers, ranchers, and lumbermen; and protecting peaceful tribes from depredations by whites.

As early as 1849, General Bennet Riley, commander of the Pacific Coast's 10th Military Department and military governor of California, recognized the need for someone to manage the large amount of ordnance that had accumulated at San Francisco, Monterey, and San Diego. Twice he petitioned the adjutant general to appoint an experienced ordnance officer to his department. In response to Riley's requests, the War Department finally directed twenty-six-year-old Brevet Captain Charles P. Stone to take a small detachment of enlisted men, along with stores and supplies, and, "take charge of the interests and business of the ordnance department in the Pacific division," and, under the orders of the commanding general there, "to take measures for properly supplying the troops in Oregon and California." His instructions from Chief of Ordnance general George Talcott also gave him broad leeway in carrying out this mission. Stone was to use his "own professional judgment and professional information" in the site selection and development of an ordnance depot as well as in other duties. He was, after all, a West Point–trained soldier and engineer, expected to be capable of handling on his own almost any military or civil situation that might arise.[5]

In 1850, the quickest way to California from the East Coast was by steamer to the Isthmus of Panama, overland to the Pacific, and then by sea once again to San Francisco. Generally, this journey took between five and seven weeks, but Stone and his command would be transporting a considerable amount of ordnance stores as well. At the time, these consisted of:

> Cannon, howitzers, mortars, cannon balls, shot, and shells, for the land service . . . gun carriages, mortar beds, caissons, and traveling forges, with their equipments . . . apparatus and machines required for the service and maneuvers of artillery . . . with the materials for their construction, preservation, and repair. Also . . . small arms, side arms, and accoutrements, for the

artillery, cavalry, infantry, and riflemen . . . ammunition for ordnance and small arms . . . and . . . stores for the service of the various arms; materials for the construction and repair of Ordnance
buildings; utensils and stores for laboratories, including standard
weights, gauges and measures; and all other tools and utensils
required for the performance of Ordnance duty.[6]

This extensive cargo required that they take the much longer
voyage around Cape Horn. On December 23, 1850, the army executed a contract with shipowner John McGaw to charter his ship
the *Helen McGaw* to transport Stone, a company of twenty-seven
soldiers, and the ordnance stores from Virginia to California. A
three-masted, full-rigged sailing ship, the *Helen McGaw* had been
built at Medford, Massachusetts, in 1847. She was by no means one
of the swift clippers that whisked passengers and freight from the
East Coast around the Horn to California in as little as ninety days.
Rather, the *Helen McGaw* was a smaller, slower vessel of almost six
hundred tons that had been employed as a transatlantic immigrant
packet by the Black Ball line on the New York or Philadelphia and
Liverpool runs. After taking on cargo from the Watervliet and
Harper's Ferry arsenals, she sailed for Norfolk, Virginia, to load
more supplies and meet Stone and his men. His command included specialists, such as carriage and wagon makers, armorers, and
artificers, all of whom would be needed at the new depot. They
had been drawn from Fortress Monroe as well as other eastern arsenals. Accompanying the ordnance detachment was Second
Lieutenant George H. Paige of the Second Infantry Regiment on
his way to "frontier duty" at Fort Yuma, California. By February 2,
1851, they had boarded the *Helen McGaw* and soon were at sea.[7]

Driven by fair winds, the *Helen McGaw* made good time and the
voyage went smoothly, with Stone's landlubbers acquiring their sea
legs and learning the routines of life on the ocean. With the crew
handling the sails and rigging and the soldiers taking care of the
deck, they crossed the equator and sailed along the South
American coast. The troops had to bathe regularly and keep their
bunks and clothes clean. They were, however, instructed not to
hang their laundry to dry on the rigging, as this could interfere with
the handling of the ship. They kept themselves amused by reading, playing cards, and, likely, singing. Stone may also have
engaged in some fishing from the quarterdeck as he had on the

Tahmaroo. The highlight of the southbound trip came with the crossing of the equator, celebrated with appropriate hazing (and shaving) of the enlisted men by King Neptune and his "wife." As William Tecumseh Sherman recalled on a similar voyage four years earlier, however, naval etiquette exempted officers from the hijinx, and Neptune was not allowed to come aft of the mizzenmast.[8]

On March 28, the *Helen McGaw* reached her first port of call, Rio de Janiero, Brazil. There she spent three days replenishing stocks of food and water. A visitor to Rio a couple of years earlier described the entrancing experience of entering the harbor:

> Terrace upon terrace, dressed in loveliest verdure; islet upon islet, resting in the deep shade of overhanging hills; turreted fortresses, spires and domes; hamlets peeping through groves of orange trees, crafts of novel structure with the graceful lateen sail, scudding swiftly by; stately ships of war with the colors of every nation at their peak, and the city rambling o'er valley and hill, formed a coup d'oeil which it does not fall to the lot of man to gaze upon—too lovely as a whole for the eye, yet unsatisfied, to attempt to discriminate its individual beauties. With shortened sail we glided slowly to an anchorage. Cheer after cheer arising from the California bound ships as we slowly picked our way among them, when suddenly luffing in the wind, we let go our anchor. . . . About a league from us lay he town of Rio de Janiero, and few cities show to better advantage. Built without regard to regularity, running along the beach, up the sides of hills, through ravines, its suburbs stretching away, circling around coves and capes . . . Abounding in churches, convents and fortifications . . . with a high mountain range for a background . . . at once picturesque and imposing.[9]

In all likelihood, Stone and Paige went ashore, perhaps, as Sherman had, attending the theater, taking in the botanic gardens, almost certainly climbing the *Corcovado* peak, with its panoramic views of Rio and vicinity, or visiting the shopping districts where attractive young women produced beautiful feather flowers—"as much to see the pretty girls as the flowers."[10] But time in what must have seemed a tropical paradise was short, and they were too soon back at sea. After taking on the supplies, which included fresh fruit and vegetables, much to the delight of everyone on board, the *Helen McGaw* continued south, where the weather got colder and the seas noticeably rougher. The soldiers also began to grow restive due to the cramped quarters and forced inactivity. To a great extent, Stone's even-tempered demeanor kept a lid on the situation. On April 25, however, he noted in the company log that an artificer had to be disciplined for disrespect to his sergeant. The unfortunate offender was sentenced to eight days' confinement, during which he had to walk, under guard, between the mainmast and the capstan on the bow continuously from nine A.M. to noon and six to ten in the evening. When not relaxing on the quarter-deck or in the ship's saloon, or sorting out disputes among his men, Stone busied himself with inspecting the ordnance stores and sitting in on a number of boards of inquiry into the healthfulness of the food and other issues regarding conditions onboard.[11]

The *Helen McGaw* made an unusually wet, windy, and uncomfortable passage around Cape Horn, but as she proceeded into the Pacific and northward along the Chilean coast, the seas calmed and the weather warmed. The monotony of the journey was broken by a celebration on July 4, when all duty, except guarding the stores, was suspended so the soldiers and sailors could commemorate Independence Day. Festivities included firing a thirty-one-gun salute (one shot for each state in the Union) and culminated with fireworks over the ocean. They had put in at the Chilean port of Valparaiso for supplies, but a yellow fever epidemic there prevented Stone or anyone else from going ashore. Stocks of food and water had to be lightered out to the ship. Then it was back to sea and on to California. On August 15, after a voyage of 173 days, the *Helen McGaw* passed through the Golden Gate and dropped anchor off San Francisco's busy wharves.[12]

Straightaway, Stone paid a call on Brevet Brigadier General Ethan Allen Hitchcock, Winfield Scott's inspector general during the Mexican War, who now commanded the army's Pacific Division. The launch taking him to shore had to navigate carefully around the hulks of ships abandoned by their gold-hungry crews and other vessels waiting to discharge their cargoes, but finally Stone touched shore in California. He made his way up San Francisco's dusty streets toward General Hitchcock's headquarters. The Gold Rush was still in full swing, and San Francisco was its economic and social heart. Everywhere Stone looked he saw an energetic and bustling city—merchants hawking their wares, carpenters and masons erecting new buildings (some to replace ones destroyed in May's most recent "Great Fire"), brokers trading mining shares on the street corners, and businessmen and bankers dashing about on blooded horses or in fancy carriages. It was all very exciting, and may have distracted the brevet captain temporarily from his mission. Perhaps someday military life would be behind him and he would become one of these men of affairs. By the time he reached Hitchcock's headquarters, however, Stone was all business, meeting with the general to present his orders and discuss locations for the ordnance depot.

Hitchcock emphasized to Stone the pressing need for an ordnance facility, and that he should find a place at which he could construct a temporary depot—one that could meet the immediate requirements and later be made permanent. Stone went to work at once evaluating potential sites. San Francisco was an obvious spot; it was the state's population center, with access to water transportation, abundant infrastructure, and building materials and labor. On the other hand, its damp weather mitigated against storing gunpowder there, warehouse rents were high, and, of more immediate concern, the city could mount no credible defense should a foreign power attack. The army had plans underway to protect San Francisco Bay, but no one knew when they would come to fruition. Next, locations at Monterey, San Diego, and ports north of San Francisco were considered, but these, too, had unfavorable climates and lacked adequate defenses. With sites along the coast ruled out, Stone turned his attention inland, and an obvious choice soon emerged. In fact, before he set sail, the chief of ordnance had suggested to him Benicia, situated some thirty miles northeast of San Francisco on Suisun Bay near the confluence of the Sacramento and San Joaquin rivers. The army already had a reservation there with barracks for infantry, artillery, and dragoons, as well as quartermaster and commissary depots. Benicia was close enough to furnish ordnance to the San Francisco Bay Area, if needed, while its inland location made it less vulnerable to attack. It could also handle deepwater vessels and more easily supply posts in the interior of the state. After weighing these advantages, Stone recommended the site to General Hitchcock, who readily agreed. Stone returned to the *Helen McGaw*, and on August 19, the ship weighed anchor for the trip across San Francisco and San Pablo bays and up the Carquinez Strait to Benicia.[13]

Upon arrival at Benicia, the ordnance detachment debarked and settled into the barracks while Stone set about finding the best location for his depot. A cool wind from San Francisco Bay coming up the Carquinez Strait may have greeted him. Indeed, after months at sea aboard the cramped and stuffy *Helen McGaw*, Benicia's open hills seemed to be very favorable. A soldier stationed there in the 1850s said of the site:

> In Benicia the climate is glorious. It seldom gets too warm. The ocean breeze makes it pleasant. The fogs so prevalent in San Francisco seldom come this far. There is no rain during the summer months. There are but two seasons here, the dry and the rainy.[14]

As he clambered up and down the treeless, dry, rolling countryside, however, Stone came to some sobering realizations. First, although Benicia was a deepwater port, the army's reservation had inadequate facilities for unloading and loading of ships. The hulk of an old French sailing vessel, the *Julie*, served as the only dock. More disturbing was the fact that the existing barracks and depots took up most of the level terrain. If the army could not expand the reservation, he would have difficulty building the ordnance depot among the hills. Those hills also were parched and prone to fires during the dry season, which lasted from around May to November, while the only sources of water were Suisun Bay and a small windmill. Nevertheless, Stone envisioned a depot with a permanent wharf and substantial fireproof buildings. To put the buildings on level ground, he requested that the army acquire an additional 393 acres adjoining the reservation. General Hitchcock agreed, and made a deal with the owners to buy the land. The secretary of war, however, disapproved of the purchase partly due to cost and partly because the proposed acquisition, as well as the reservation itself, was part of Mariano Vallejo's Rancho Suscol Mexican land grant, the legal status of which remained unclear for some time. Hitchcock, therefore, was compelled to use his personal funds to buy the property, which he then resold to private parties. Stone would have to place the depot—at least the "temporary" depot—in the hilly portion of the reservation, although he did start lobbying through the chief of ordnance to relocate the quartermaster and commissary so that the permanent facility could be moved to their sites.[15]

When the *Helen McGaw* arrived off Benicia, Stone found the decks of the *Julie* already crowded with quartermaster and commissary goods. He could not begin to unload his ship for several days. Then there was the matter of where to put the stores once they were brought ashore. Stone had selected a location, but the hills in and around the reservation were bare, and competition from booming San Francisco–made lumber was expensive, as were bricks and civilian carpenters and masons. He decided, therefore, to forego brick construction, buy the lumber locally as cheaply as he could, and have his own men build the depot. The work of the company was made more difficult by the desertion in early September of an armorer, carriage maker, and two other soldiers who saw their sea voyage as a free ticket to the gold fields. Nevertheless, construction went ahead, and by the end of the month a forty-by-twenty-foot frame magazine had been completed. The impending onset of the rainy season made this building a necessity to keep supplies of gunpowder dry. Other temporary wooden structures followed in 1851 and 1852, including a 140-by-28-foot building housing the carriage maker's, blacksmith's, armorer's, saddler's, and painter's shops, quarters for twenty men, a storehouse and gun carriage shed, and a stable. Because of delays at the War Department in processing requests for funds for these projects, Stone used his own money to pay for some of the materials, hoping for later reimbursement.[16]

It does not appear as though the army intended Stone to be the permanent commander of the ordnance depot. In October 1851, Colonel Rufus L. Baker, who had charge of the Watervliet Arsenal, was ordered to the West Coast. Baker, however, felt that his health would suffer at the new assignment. He refused to go and resigned from the army. As a result, Stone remained in command of the ordnance depot, although its development was not his only responsibility. Soon after arriving in California, he began receiving information that the Pacific Department's existing ordnance stores were in disarray. Stone and his senior enlisted personnel then launched a series of investigations. Beginning in the San Francisco Bay Area, he determined large amounts of weapons and supplies to be useless due to improper storage. At Monterey, he and his men found the redoubt that housed muskets, gunpowder, and ammunition to be unguarded. The post adjutant's office was deserted as well, making it a simple matter for anyone to pilfer weapons and other stores. Subsequent trips to posts as far away as San Diego turned up similar conditions, and reports of many substandard rifles and muskets in Oregon also reached Benicia. Stone definitely had his work cut out for him. He had little time to mourn the passing of his father on September 5, 1851, or his mother that November.[17]

Stone dealt with the Oregon weapons problem quickly, issuing orders for all rifles and muskets (with the exception of those in the best shape) to be sent to Benicia for reconditioning. Likewise, he directed all army posts in California to send their unused or damaged guns and equipment to the new depot for repair and safekeeping. The arrival of such large amounts of ordnance and stores from West Coast locations as well as from Eastern arsenals made the construction of permanent facilities imperative. As early as January 1852, he had requested from the chief of ordnance funds for a new fireproof magazine. Because of the high price and poor quality of local bricks and mortar, Stone suggested that these materials be purchased in the east and shipped to Benicia. He also asked for funds to build a wharf to replace the rotting *Julie*. While these requests wound their way through the army's bureaucracy, weapons and stores began arriving at Benicia from up and down the West Coast. The powder issue was solved temporarily by placing the excess barrels aboard a schooner anchored in Suisun Bay for the dry (fire) season and then moving them ashore under what cover could be found once the fall and winter rains began. Other stores were kept under tarps or in lean-to sheds. Such expedients, however, could not be expected to continue very long, and still Stone heard nothing about his requests. Finally, frustrated with Washington's refusal to build adequate facilities at Benicia, in 1853 he took matters into his own hands and initiated construction of a permanent ordnance wharf, a large, two-story storehouse, and a small engine house. The buildings were located on the site of the temporary depot since pleas by Stone and the Ordnance Department for an expansion of the military reservation or relocation of the quartermaster and commissary depots also went unheeded by the War Department. The new storehouse was not made of brick, but of fine sandstone quarried from deposits on the reservation. This technique did not comply exactly with army practice, which generally favored brick, but it met Benicia's pressing needs in a creative and economical manner. Once again, Stone financed some of the new storehouse himself.[18]

By the time construction began on the permanent buildings, ordnance work at Benicia was already in high gear. In 1852, the depot had fabricated over forty thousand cartridges for small arms and three hundred pounds of bullets for Colt's revolvers, made serviceable sixteen artillery carriages and caissons, 590 small arms, and six thousand rifle cartridges. The staff cleaned, oiled, and repacked 1,500 cartridge boxes, and removed from their cases, oiled, and repacked over 1,600 muskets, rifles, and carbines, as well as 380 swords and sabers. Stone's command also relocated nine thirty-two-pounder garrison guns to the old Castillo de San Joaquin at the entrance to San Francisco Bay. The horseshoe-shaped adobe Castillo had been built by the Spanish in 1794 but had fallen into disrepair and was, to a great extent, a ruin by the time Stone arrived. The installation of guns at Castillo de San Joaquin may be related to an incident later recounted by Stone's West Point classmate, Fitz-John Porter. According to Porter, during the 1850s the French consul general in San Francisco perceived an insult to either his nation or himself and threatened to have the French fleet seek satisfaction. The commander of the Pacific Division charged Stone with mounting some kind of defense against this possibility. Among his preparations was the location of batteries at the entrance to the bay, along with a couple of mobile guns with shot-heating equipment for use against wooden-hulled ships. Eventually, the consul calmed down and the crisis passed without further incident. Whatever the reason, Stone's placement of the thirty-two-pounders provided San Francisco with its first significant protection from hostile vessels during the American period. The growing importance and accomplishments of the ordnance depot led to its designation in 1852 as the Benicia Arsenal, placing it on a par, at least administratively, with similar, well-established facilities at Watervliet, Allegheny, near Pittsburgh, Washington, DC, and Saint Louis.[19]

Over the summer of 1853, Stone was at Fort Vancouver on the Columbia River assisting in the establishment of an ordnance depot there. While at Fort Vancouver, he also worked with Brevet Captain George B. McClellan of the Corps of Engineers in outfitting the survey of the most northerly of the Pacific Railroad routes, and renewed his acquaintance with the post's quartermaster, Lieutenant Ulysses S. Grant, who had taken part in the first attempt to summit Mount Popocaptetl five years earlier and had visited Benicia briefly before going to the Pacific Northwest. From Fort Vancouver, Stone moved on to southern California, where he spent the end of the year and the first part of 1854. At San Diego, he equipped the Topographical Engineers as they set out to survey the southern Pacific Railroad route across the Southwest under Lieutenants John Parke and John Pope. During these and other prolonged absences, Stone relied on his second in command to keep the arsenal functioning. Military Store Keeper William Whitridge held this responsibility until he went absent without leave in February 1853 and had to be replaced. Four months later, Lieutenant William Weckler of the Ordnance Department arrived at Benicia. Over the next three years, he would serve as Stone's assistant and aide-de-camp.[20]

Whether Stone or Weckler was in command, life at the Benicia Arsenal revolved around a standard military routine. A three-man detail guarded the powder magazine twenty-four hours a day. For those not on guard duty, drum cadences awoke the company and sent the men to breakfast at 6:30 in the morning. Then it was off to work, either doing construction of facilities or ordnance assignments. Some of these tasks were described by a civilian visitor in 1855:

> Soldiers do the work making gun carriages, stocking guns, and repairing all the most finished work of heavy ordnance; infantry rifles, and cavalry accoutrements are prepared, cleaned, kept in order, and made ready for shipment to all portions of the Pacific Coast.[21]

Dinner came at noon, followed by more work from one to five P.M., when release was sounded. Retreat was sounded at sunset, with taps being beaten at ten o'clock at night. On Sundays, the officers conducted an inspection of the post at nine A.M. Occasionally, drills and parades were also held. All in all, conditions at the arsenal were not nearly as rigorous as at many other western posts, and Stone, while he expected discipline to be maintained and his orders to be carried out, was not a particularly inflexible or harsh commander—on one occasion he publicly withdrew a rebuke given to two enlisted men for an infraction after learning that they were not guilty. Nevertheless, unauthorized absences and desertions proved to be a constant worry. The treasures said to be waiting in California's nearby Mother Lode country prompted many to desert, as did the low wages paid to enlisted personnel. William T. Sherman recognized this when he recalled:

> I have seen a detailed soldier who got only his monthly pay of eight dollars a month and twenty cents a day for extra duty nailing on weather boards and shingles along side a citizen who was paid sixteen dollars a day. This was a real injustice, made the soldiers discontented, and it was hardly to be wondered that so many deserted.[22]

A deserter could easily lose himself in the transient and largely male population of the gold fields, where he might even be sheltered by sympathetic miners, or in the teeming city of San Francisco. Although capture meant severe punishment including loss of rank and imprisonment, the risk seemed worth it to soldiers who found army life too onerous. Other dissatisfied troopers sought to ease their troubles with alcohol. Stone usually dealt with such overindulgence by having the transgressor walk the post from six to ten P.M. for seven days while carrying a knapsack loaded with three nine-pound shot, or, in more severe cases, ordering confinement to the arsenal for up to six months.[23]

Problems with isolation at the Benicia Arsenal were mitigated somewhat by the presence of the barracks and quartermaster and commissary depots on the reservation. Soldiers from the facilities mingled and enjoyed leisure-time activities together. The arrival of Captain Robert E. Clary to be the quartermaster at Benicia brought a pleasant change to Stone's life of duty and service. A fellow native of Massachusetts, Clary was a career army officer and graduate of the West Point class of 1828. Classmate Jefferson Davis, now the secretary of war, had served as the best man at his wedding to Esther Phillipson, a member of a prominent Saint Louis mercantile family. Accompanying Clary to California were his wife and family, including an attractive daughter, twenty-year-old Maria Louisa. Soon, Stone was smitten and paying many visits to the quartermaster's residence courting the young woman. The attraction was mutual, and the two married in Benicia on July 5, 1853. In

October of the following year the couple welcomed their first child, a girl named Esther, for her grandmother, but whom they affectionately referred to as "Hettie."[24]

By 1854, the Benicia Arsenal had a complement of two officers (Stone and Weckler) and forty-four enlisted men, along with a handful of civilian employees. This fairly small staff maintained a wide array of ordnance stores, which included over twelve thousand percussion muskets, 5,700 percussion rifles, 1,000 musketoons, 200 carbines, 1,190 percussion pistols, 280 Colt's revolvers, eleven six-pounder cannon, six twelve-pounder brass howitzers, forty-seven coast defense guns (twenty-four- and thirty-two-pounders), three mortars, and 1,690 sabers and swords, along with gunpowder, appropriate ammunition, forges and wagons, and limbers, carriages, and caissons for the artillery. From these stores, Benicia provided ordnance and maintenance for the army's new bastion at the Golden Gate, Fort Point, on which engineers had begun construction in 1853; for campaigns against the Native Americans throughout California, Oregon, and Washington; and for state and territorial militias, as well.[25]

The growing importance of the arsenal caused the commander of the Pacific Department, Major General John E. Wool, to order the building of a second permanent sandstone storehouse, "without delay," in March 1854. The directions received by Stone set the cost of the structure at no more than twelve thousand dollars, and allowed him to draw from the Ordnance Department "from time to time . . . such sums as he may require for the fulfilment of the above order."[26] Despite this assurance, requests for payment were delayed due the fact that Stone had begun work on the building upon the orders of his commanding general who had not obtained congressional approval for it. Congress protested further payment of drafts Stone submitted and they were held up indefinitely. The contractor, however, had already been paid a quarter of the money and insisted on completing the storehouse before the rainy season started. Stone, caught in the middle, also had his salary suspended and was compelled to obtain a loan of over nine thousand dollars, plus interest, to cover the ongoing costs of construction and his own expenses. This shameful situation was finally resolved with the discovery of an old law providing that junior officers could not be held monetarily liable when operating under the orders of their commanders. When brought to light later that year, the law enabled Stone to recoup his back pay, and a congressional appropriation funded completion of the storehouse (at a final cost considerably above the stipulated amount) as well as other permanent arsenal buildings, permitting him to repay the loan. Nevertheless, the whole affair left Stone embarrassed and likely starting to question the financial wisdom of remaining in the army.[27]

In July 1854, Colonel Joseph K. F. Mansfeld visited Benicia on an inspection tour of the Department of the Pacific for the secretary of war. He applauded Stone's leadership of the arsenal, particularly his selflessness in fronting the money for construction of the second permanent storehouse. Nevertheless, Mansfeld found the lack of a fireproof powder magazine disturbing. The current frame magazine was buckling under the weight of the powder housed therein, and, despite the twenty-four-hour guard, it was susceptible to fire, either from sabotage or carelessness. He knew Stone had pushed for a permanent building as part of his plan for the arsenal, and hoped that the government would soon provide the funds for one. Despite Stone's urging and Mansfeld's recommendation, however, support for a new magazine was not forthcoming from Washington. When a writer from the *California Farmer* journal visited the post in the spring of the following year, he was aghast that some twelve thousand pounds of gunpowder were still stored in a wooden building. Were it to catch fire, he speculated that the resulting explosion would destroy the arsenal and other military facilities on the reservation, as well as the town of Benicia. Even ships docked at its wharves would be in danger. He, too, knew that there were plans to rectify the deficiency, but a lack of funding kept the project in abeyance. The writer sympathized with Stone's situation and noted that the lieutenant was becoming weary of having all the responsibility but little support for his management of the arsenal, commenting with tongue in cheek: "Constant care and watching . . . will wear away a Stone."[28]

When funding did finally become available, Stone initiated construction of a new brick magazine building. By this time, however, much of his attention was elsewhere. He had been promoted to first lieutenant, but still found it increasingly difficult to support his family on a junior officer's salary, let alone having to deal with the financial pressures the second warehouse debacle had placed on him. Goods and services on the West Coast commanded exorbitant prices, and the pay differential the military provided for soldiers there did not go far enough. As Ulysses S. Grant famously put it, "A cook could not be hired for the pay of a captain. The cook could do better."[29] As a result, beginning during the Gold Rush the army encouraged officers, particularly junior officers, to supplement their incomes with private-sector work, and many did just that. William T. Sherman engaged in surveying for ranchers and acquired land that he later sold at a profit. Grant tried his hand at raising potatoes and selling pigs and chickens. On a much grander scale, Captain Joseph Folsom, quartermaster at San Francisco, invested in city lots, becoming by the time of his death in 1855 one of the richest men in the state. Likewise, Captain Henry Wager Halleck, who had served as the army's secretary of state for California prior to statehood and played a key role in drafting its 1849 constitution, practiced land law in San Francisco and became the senior member of the firm of Halleck, Peachy, and Billings, one of the most prominent legal firms of the day. If their private work resulted in conflicts of interest with their military assignments, most officers got off with only a reprimand.[30]

Following the examples of Folsom, Halleck, and other officers, Stone entered the world of private enterprise. As early as 1854 he had been dabbling in gold and bullion trading—purchasing dust or nuggets from miners, refining them into bars, and selling them to banks and mints either in San Francisco or the East at a profit. His office was in the city on the corner of Jackson and Montgomery Street, the heart of the financial district at the time. He achieved some success in this endeavor, and a year later joined in the formation of the San Francisco and Sacramento Railroad Company. The line was to have run from Benicia through Solano and Yolo counties, to the west bank of the Sacramento River opposite the city of the same name. Stone was one of the leading stockholders in the firm, purchasing fifty shares at a hundred dollars per share, second only to pioneer railroad engineer Theodore Dehone Judah, who owned ninety shares. Other investors included the former American consul to Mexican California, Thomas O. Larkin, now a San Francisco entrepreneur, and Mexico's ex-military governor,

Mariano G. Vallejo. Stone became secretary of the company, but despite the involvement of such prominent investors, the project could not raise sufficient capital and never advanced beyond a preliminary survey. For all his hard work, Stone received no returns from this investment.[31]

While participating in the railroad scheme, Stone also became involved in the volatile world of California banking. The state's constitution had prohibited the establishment of traditional banks, but this did not stop merchants and other businessmen from making loans and taking deposits without actually calling themselves banks. Within a few years, a number of such companies operated, primarily in San Francisco, but with subsidiaries and partners throughout the gold region. One such enterprise was the prominent Saint Louis banking house of James H. Lucas and Henry S. Turner, which opened a San Francisco office in 1853. Heading this branch was a junior partner, William T. Sherman, who had earlier gone east, married, resigned from the army, and joined the firm. Not long after his return to the West Coast, Sherman, one of the city's leading businessmen, became acquainted with Stone. Just how they got to know one another is unclear, but they were both West Point graduates, and Stone's bullion trading office was located at the same intersection as the Lucas, Turner & Co. building. Within a short time they had become close associates, if not friends. Sherman developed into a sort of mentor to Stone, leading him into the world of business and showing him the ins and outs of California banking.

Evidently, he was impressed with his protégé. In early 1855, Sherman fell ill with a bout of asthma (and not a little hypochondria). Fearing for his life, he wrote his Saint Louis partners that, in the event of his incapacity or death, he nominated Stone to be his successor, describing him as "the best man I know, competent and willing to enter on the task."[32] Sherman did not pass away, though by the end of the year he remained weak and pale, still feeling as though he "might die at any moment," and recommending that in that event, Stone should take over for him and assume his one-eighth share in Lucas, Turner & Co. He even stated that he would leave the lieutenant his power of attorney, and tried to convince him to purchase his San Francisco house and furnishings should his condition worsen.[33]

Sherman's confidence in Stone must have boosted the latter's self-esteem and pushed him in the direction of greater involvement in banking. The profits to be made in the private sector clearly outweighed his army salary, and his duties as ordnance officer for the Pacific Division caused him to be increasingly absent from his family. As 1856 progressed and the second storehouse at the arsenal rose, Stone found himself taking leaves to devote more and more of his time to business. He moved Maria Louisa, who was now pregnant with their second child, and Hettie from quarters on the reservation to a rented house in San Francisco on Vallejo Street above Stockton. His company, known as C. P. Stone, entered into a partnership with longtime expressman and banker Sam Langton of the gold mining town of Downieville, in the northern foothills of the Sierra Nevada, and businessman N. N. Wilkinson of Marysville, an inland port near the confluence of the Yuba and Feather rivers. The new firm, known as Langton's Pioneer Express, proposed "transacting of a general Banking, Exchange and Gold Dust business." The firm also had offices in the mining camps of Nevada City and Forest City, and carried on the existing shipping of freight and "treasure with messengers operating between Downieville and San Francisco."[34]

With Langton taking care of the business in Downieville and the gold country and Wilkinson seeing to river shipments from Marysville, Stone managed the San Francisco office. It was quite similar to the work he had been doing as a bullion trader. He received the dust and nuggets purchased from the miners, and had the gold refined and made into coins and bars for investment or shipment to the East Coast. To carry out this operation, Langton's Pioneer Express required a considerable amount of capital, which had to be borrowed from San Francisco banks. Stone used his relationship with Sherman to obtain a $25,000 loan from Lucas, Turner & Co., which he sent on to Langton for gold purchases and loans to miners. To check on his investment, Sherman joined with Stone in October 1856 on an inspection trip to Langton's "upcountry" offices. While they found the books in order and the company, on the whole, running profitably, Sherman had misgivings about some of the loans Langton had made. In particular, he questioned the soundness of advancing money to a placer mining operation at Bidwell's Bar on the south fork of the Feather River. It consisted of a wooden dam to divert the river and dry up its original bed, thereby, hopefully, exposing gold deposits. Langton had plunged $13,000 into this scheme, which Sherman thought to be "a grand gamble . . . and . . . unsafe." He instructed Stone not to apply any of the funds obtained from Lucas, Turner to the venture. While Sherman's military nature admired the "skill and boldness" of the miners, his banking sense saw their projects as far too risky. He perceived danger in Langton's policy of issuing checks to miners in anticipation of their coming up with the gold to be sent to San Francisco, and urged Sam Langton to close some of his Sierra foothills branches that seemed to be playing a bit fast and loose with the firm's money. Sherman confided these observations to Stone, who agreed not to continue funding any of Langton's speculations and eventually ended his partnership with him.[35]

Sherman's doubts about Langton's Pioneer Express proved prescient. Langton was, by all accounts, quite overextended. Wilkinson also sensed something was amiss and withdrew from the firm. By December 1856, banks in San Francisco had begun refusing Langton's checks, causing runs on the company's offices in Downieville, Nevada City, and Forest City that it could not cover. The collapse of Langton's Pioneer Express had an impact on Stone as well. He had not gotten out of the business soon enough, and still had a considerable amount of money with Langton—both his own and that of fellow soldiers whom he had convinced to let him invest for them. One of the largest of these accounts came from his father-in-law, Captain Robert Clary. He and his wife had borrowed $20,000, which they turned over to Stone for investment. Stone had put some of the money entrusted to him with Langton, but much went into real estate and other tangible assets. Nevertheless, it would take some time for him to convert these into the cash needed to repay his investors and deal with the losses Langton had caused. Time, however, was something Stone did not have. Most of his investors clamored to be paid right away. The final blows came when Stone went up to Downieville to attach Langton's property. At the same time, his clerk in San Francisco disappeared, and a closer examination of the company's books showed that the clerk had embezzled up to $40,000 to support his mistress and a gambling habit. Stone had seen that account declining, but assumed the funds were being withdrawn by Sam Langton for investment purposes. Although Stone hurried back to the city when he heard about the theft, leaving a Marysville attorney to deal with Langton, there was little he could do. The doctored accounts also showed that Stone was owed only about half of the amount he thought Langton owed him. Stone realized he was ruined. It would not be the last time, however, that betrayal by an associate and inattention to the actions of a subordinate would get him in trouble.[36]

As 1857 approached, Sherman described Stone as "completely disheartened and dejected," and afraid to face him and tell him the truth. His opinion of him changed abruptly. As late as October 1856, Sherman had still believed Stone to be an honest hard worker and could do equally as well as he, should the lieutenant be called upon to succeed him at Lucas, Turner. Nevertheless, Stone's inability to rein in Langton's speculations proved troubling, and within a couple of months Sherman had lost confidence in him altogether, finding it "beyond comprehension" that he could not have known what was going on at Langton's Pioneer Express. He withdrew his recommendation for Stone's succession, favoring either a "Jew . . . without feeling, who is not to be moved by the appeals of a man in imminent peril of ruin," or an older, more experienced banker. Sherman recorded that he felt very bad about Stone's troubles, but, noting that "business is business," resolved to have nothing more to do with him.[37]

Facing the prospect of financial ruin and dishonor for losing his comrades' investments, Stone felt he could no longer remain in the army. Perhaps he also recalled the phrase from the old West Point drinking song, "Benny Havens, Oh!"—"In the army there's sobriety, promotion's very slow"—and realized that he would probably stay a junior officer, with a junior officer's meager salary, for some time. Even if he made captain soon, in a peacetime military he could expect to remain in that rank for another twenty years before being promoted. He resigned his commission in November 1856, intending to dedicate all of his time and effort to supporting his family (which on December 1 included baby Charles Jr.) and trying to make good on his mounting debts. The attachment against Langton resulted in Stone receiving only offices, furniture, mules, and wagons, all of little value. Over the next few months he sold what other assets he could, including a farm in Benicia, mostly at a loss, and used the money to try and repay those who had placed their trust in him. The financial climate in San Francisco, however, was not conducive to such a recovery. Stone was among the 271 insolvencies in the city during the period from 1856 to 1857, and many of these were scrambling to sell what they could as well. Even Sherman commented that, "Nearly all the rich men of 1853 are bankrupt." Under California law at the time, unless Stone had knowingly committed fraud, which he had not, his debts could be quickly and legally cancelled by the court. But Stone's sense of personal honor compelled him to do whatever he could to meet his obligations and recoup his reputation. In this he achieved some success, raising enough cash to reimburse most of his investors. While this may have raised Stone's spirits somewhat, his world once again came crashing down with the death of Charles Jr. on April 17, 1857, at the age of four months and seventeen days.[38]

Perhaps it was sympathy for the passing of Charles Jr., or Stone's honest and earnest attempt to redeem himself, that eventually softened Sherman's opinion of him. In 1857, however, Sherman initiated the closure of the San Francisco branch of Lucas, Turner and headed east to manage the firm's New York office. Between 1854 and 1857 Sherman had guided Stone through the intricate world of California banking. Now Stone could no longer rely on his help and advice. During this time, Sherman formed an impression of the brevet captain as genial, energetic, faithful, and honest, if somewhat naive. It is likely that these traits would be in the back of Sherman's mind a decade and a half later, when he again stepped in and changed the direction of Stone's life.[39]

In the meantime, despite grieving over the death of his son, Stone tried doggedly to find some form of civilian employment that would support his family. This was not an easy task, for, as Sherman pointed out before leaving San Francisco, "California is no place to recuperate in," and thought that the disgraced banker might do better by returning to the army.[40] That was something Stone felt he could not do, given the low pay and his embarrassment resulting from the Langton debacle. The nationwide depression that followed the Panic of 1857 was already beginning to be felt on the West Coast. The boom times of the Gold Rush had also started to fade as the easily accessible mineral deposits played out, and the discovery of Nevada's Comstock Lode, which would revive San Francisco's business community, was still in the offing. For now rejection followed rejection. Just as his situation seemed desperate, however, a sea captain associated with a Mexican banking and mercantile house, which Stone would later find out had a checkered past, approached him with a proposition that just might turn his fortunes around.

MEXICAN MISADVENTURE

ANKER Jean Baptiste Jecker hailed from a part of Switzerland that in the early nineteenth century was claimed by France. As a young man, he accompanied his brothers to Mexico in 1835, where he learned well the businesses of international trade and banking. His influence grew and he began arranging foreign loans to the cash-starved, newly independent government at usurious rates. When the loans could not be repaid, he was given control of various state functions such as the collection of taxes, customs duties, and mail services. While such arrangements were scorned by patriotic Mexicans, they made Jecker wealthy and one of the country's most prominent financiers. He joined with a Spaniard, Isidoro de la Torre, in the 1840s to form the commercial house of Jecker, Torre & Co., with headquarters in Mexico City. Its activities soon expanded to Mexico's West Coast ports of Mazatlán in the state of Sinaloa and Guaymas in Sonora. As California boomed during the Gold Rush, Jecker, Torre also opened an office in San Francisco. In August 1856, the firm entered into an agreement with the Mexican central government to survey, map, and describe the "*terrenos balidos*," described as "all the property of the Federal Government, waste lands, the old presidios, the Jesuit and

Franciscan Missions, the lands of barbarous tribes of Indians, ene-mies of the white race, who have never submitted to the laws; and, lastly the lands occupied by private individuals to which they have no legal title conformable to Mexican laws" in the northern fron-tier state of Sonora.[1] There were two principal stipulations: First, the work would have to be done by a "scientific commission," funded by Jecker, Torre. And second, it had three years to com-plete the contract or else risk cancellation. Once the work was com-pleted and approved by Mexican authorities, Jecker, Torre would receive a whopping one-third of all the lands mapped.[2]

The survey contract was not Jecker, Torre's first Sonoran ven-ture, a fact that later would prove detrimental to the execution of the 1856 agreement. In 1850, Sonora enacted a law to promote development through colonization. It gave entrepreneurs who brought in settlers extensive land grants and exemption from taxa-tion. Although the Mexican Congress declared Sonora's action unconstitutional the following year, the act did serve to stimulate foreign interest in the remote region. One group that paid particu-lar attention to Sonora at this time consisted of Frenchmen who had come to California to find their fortunes during the Gold Rush and subsequently encountered legal and physical abuse at the hands of nativist Americans. By the early 1850s, many were ready to leave the United States if the chance presented itself. For its part, Mexico certainly did not want to encourage American settle-ment, having lost half its territory to its neighbor in the 1840s, but the French were another matter. Most were Roman Catholic, and more likely to adopt Mexican citizenship and assimilate easily into Mexican society. Seeing a lucrative opportunity in Mexico's over-tures to the French, Jecker, Torre stepped in, and in 1853 entered into a partnership known as the *Compania Restraudora de Mineral de Sonora* with an adventurer of minor French nobility living in San Francisco, Gaston de Raousett-Boulbon. The *Compania* obtained a Mexican concession for the survey and development of mines in northern Sonora, and with Jecker, Torre's money, Raousett-Boulbon recruited about two hundred French expatriates to partic-ipate in the project. He also obtained an ample supply of arms in San Francisco for the party, ostensibly to fight off hostile Indians. When the would-be colonizers arrived in Guaymas, however, they found the Sonoran authorities less than welcoming. A couple of smaller colonizing parties, consisting largely of Frenchmen, had already proven troublesome and not really interested in promoting the development of the state. Raousett-Boulbon's arrogance in dealing with local officials and the paramilitary appearance of his followers further alienated the Mexicans. When he began fomenting discontent among residents along the frontier with the United States, and appeared to be preparing to lead a possible independence movement, the Sonorans took action. They organized a strong military force, which compelled Raousett-Boulbon and his men to give up their arms and return to California. Two years later, the persistent Frenchman returned to Guaymas bearing a new agreement for the colonization of Sonora from President Santa Anna. This time he brought with him four hundred of his countrymen, once again armed to the teeth for "self-protection," and likely funded by Jecker, Torre. The Sonorans regarded this incursion as a filibustering expedition and wasted no time in defeating the interlopers and executing the hapless Raousett-Boulbon.[3]

Although Jecker, Torre's Sonoran undertaking with Raousett-Boulbon proved a disaster, the firm did not lose interest in such projects. The overthrow of Santa Anna offered another opportunity to realize riches in northwestern Mexico. Jecker quickly began negotiations with the new government which culminated in the August 1856 agreement. To carry out the contract, in early 1857 J. B. Jecker & Co. (Isidoro de la Torre having left the firm) brought in as investors Mexicans Manuel Payno, author and politician, and merchant and banker Antonio Escandon, along with a former sea captain on the Panama-California run and San Francisco businessman, Joseph Brent Griswold Isham. Each of the partners would receive a quarter of the one-third of the surveyed land promised to Jecker in the contract. Isham took the lead, agreeing to begin the survey within six months and complete it within the period specified—which was by now down to about two and a half years. Isham set about recruiting a leader and staff for what had become known as the Sonora Survey Commission. In San Francisco he approached Charles P. Stone, almost broke and unemployed, with a proposition he could not turn down.

The two men got along well and, despite Stone's recent misfortunes, Isham saw in him a trained engineer and soldier, with command experience, who had spent some time in Mexico as well. He offered the position of chief of the commission to Stone, who, lacking any other prospects, jumped at the opportunity. The new chief's duties were many. He had to organize the survey within the parameters set forth in the original Jecker and Isham agreements, establish a headquarters office in Guaymas, equip and send out survey parties, keeping them safe from raiding Apache Indians, prepare maps and charts to send to Jecker and Isham, remove and replace employees if necessary, maintain good relations with local and state officials, and, perhaps, do some surveying himself. He also had to promote the project to potential investors in San Francisco, a task in which his pleasant personality and apparent persuasiveness proved an advantage. For all of these duties, Stone received a respectable monthly salary of $600. By May 9, 1857, the staff, selected by Isham, presumably with Stone's advice and concurrence, was in place. It consisted of three engineers: Jasper S. Whiting, the brother of army lieutenant William H. C. Whiting, who graduated at the top of Stone's 1845 West Point class; Allexey Von Schmidt, who would survey California's eastern boundary in the 1870s and develop an unsuccessful scheme to ship water by tunnels from Lake Tahoe to San Francisco; and William Denton, later a prominent rancher in Baja California. There were also three assistant engineers, a geologist, a principal and assistant draftsman, a combined quartermaster, commissary, and paymaster. Other employees, such as chainmen, packers, and clerks, would be hired on site as necessary.[4]

After the commission was organized, Stone journeyed to Mexico City, where, acting under Isham's power of attorney, he negotiated amendments to the August 1856 contract. The most significant of these changes allowed Stone to subdivide up to fifty percent of the lands to be received by Jecker and his partners in order to raise funds to conduct the survey. The survey could be armed for its own security but could not occupy by force any part of Sonora. The commission had to employ at least three Mexican citizens, and a special Mexican judge would oversee its work. From Mexico City, Stone traveled to Washington, where he met with President James Buchanan, Secretary of State Lewis Cass, and Secretary of War John B. Floyd. He assured them that the survey was a legitimate undertaking, approved by the Mexican government, and not a filibustering party intending to alienate the northern portion of

Mexico. For their part, the president and the secretaries expressed satisfaction with his explanation of the arrangement, stating that it was within the bounds of international law. Stone then moved on to New York City, where he purchased scientific instruments and other supplies needed for the survey. Back on the West Coast, he set about finding investors to underwrite portions of the expedition. He was able to interest San Francisco attorneys James E. Calhoun, son of the late South Carolina statesman and politician John C. Calhoun, Peter Della Torre (no relation to Isidoro de la Torre), and Tully Wise, businessman William Lent, and pioneer orchardist William Neely Thompson. The new investors would pay $20,000 to outfit the survey and no more than $230,000 over the course of the project, and in return would receive one-half of the one-third of the lands to be charted under the original Jecker contract.[5]

With the promise of sufficient funding, on November 17, 1857, Stone sent the first contingent of the commission under Jasper Whiting to begin surveying the desolate tablelands and desert in northwest Sonora bordering the Colorado River. They traveled overland to Fort Yuma and the civilian settlement of Colorado City. There Whiting hired chainmen and other assistants and acquired a sloop, whaleboat, and skiff along with pack animals for working away from the river. Once outfitted, the party loaded aboard the boats and proceeded downriver to start the Sonora Survey.[6]

About the same time as Whiting left San Francisco, the commission as a whole gained some additional responsibilities. In Mexico City, Isham entered into an agreement with the heirs of Mexico's late emperor, Augustin de Iturbide. Several years earlier, the Mexican Congress had voted the heirs a sum equivalent to $1 million for services rendered by Iturbide during the war for independence against Spain. While the treasury lacked the ready cash for such a payment, one thing the country did have in abundance was land. A subsequent arrangement, therefore, gave the heirs nine hundred square leagues of the public domain. At first the Iturbides intended to locate their grant in Texas, but that jurisdiction gained its independence before they could complete the process. They then sought a tract of land in Alta California north of San Francisco, but that plan also became moot when the United States occupied and annexed the province. Finally, the heirs were given public lands in the states of Sinaloa and Sonora and the territory of Baja California. This time the recipients moved quickly to file on about ninety-five square leagues that were already surveyed, leaving the bulk of the concession still unclaimed due to inadequate maps. To gain a better understanding of their lands and establish title, therefore, the heirs contracted with Isham to survey and map them in return for a one-third interest in the still-unclaimed portion of the grant. Between the Jecker and Iturbide contracts, Isham now stood to become one of the largest landholders in northern Mexico. Since Stone and the commission were surveying public lands in Sonora anyway, he arranged for the Iturbide grant work to be folded into their duties.[7]

While Isham negotiated with the Iturbide heirs, Stone, back in San Francisco, learned that his investors were having second thoughts. In February 1858, Torre, Wise, and Lent sold their shares in the Jecker contract to Calhoun. This caused Stone to put aside his survey preparations and draw up another agreement between Calhoun and the cartel of Jecker, Payno, Escandon, and Isham. Calhoun became the sole potential recipient of lands that had previously been promised to his ex-partners, and agreed to take all of the responsibility for providing funds to the commission when requested. Once he had finished drafting the document, Stone returned to organizing the main body of surveyors for their duties under both the Jecker and Iturbide contracts. He acquired a 174-ton brig that had originally been in service of the United States Coast Survey. Stone renamed her *Manuel Payno*, hired an able captain and crew, and set about equipping the vessel with instruments, stores, and an ample stock of firearms. Finally, on March 16, 1858, the same day as the new Calhoun arrangement was finalized, Stone and the commission boarded their ship and sailed out of the foggy Golden Gate for Mexico.[8]

Given Stone's hectic schedule over the past twelve months, he welcomed the opportunity to be once again at sea. He must have had a feeling of satisfaction over what had been accomplished. A year ago he had been a failed and disgraced banker, without work and almost destitute. Now he was the head of a major scientific commission, staffed with some of the leading civil engineers on the West Coast. He had had the honor of explaining in person the mission of the survey to the president of the United States, who had given it his approval. Stone's income was more than sufficient to support his family back in San Francisco and pay off the final obligations resulting from his earlier insolvency. When the project was completed, he might even expect a bonus, perhaps in land or mining claims.

Aided by prevailing northwesterly winds and favorable currents, the *Manuel Payno* made a swift passage south into warmer Mexican waters. Stone may have fished from the small quarterdeck, but more likely spent his time refining plans for the upcoming survey. Before reaching Sonora, the *Manuel Payno* sailed to Sinaloa and put in at the port of Mazatlán. There Antonio Maria Vizcayno, the special judge appointed by the Mexican government, joined the expedition. He would monitor the work of the Americans, examine and approve the maps they produced. Since his salary and expenses were being paid by the commission, he could also be counted on to resolve any issues that might arise with local jurisdictions. Before leaving Mazatlán, both Vizcayno and Stone received copies of orders from Mexico City instructing the Sonoran civil and military authorities to provide the surveyors every assistance possible. With these orders and Judge Vizcayno to back him up, Stone felt more confident than ever that Sonora would mark a turning point in his life. It would, but not in the way he anticipated.[9]

In early April 1858, the *Manuel Payno* pushed her way up Mexico's western coast and into the Gulf of California. On the thirteenth she entered Guaymas Harbor, saluted the Mexican flag, and dropped anchor. As the principal port of entry, Guaymas occupied a place of importance in the state of Sonora. The harbor was practically landlocked, protected from high winds and rough seas by barren desert hills. The town itself sat at the base of one of these hills and boasted a population of between two and a half and three thousand. It consisted mostly of one-story adobe buildings, with only a few newer brick structures scattered here and there. Along the bay were the mole, the customs house (the state's principal source of income), and the more important commercial firms. Once ashore, each member of the survey recorded his name at the customs house, and Stone registered his vessel. He announced his plans, introduced Judge Vizcayno, presented the orders from Mexico City, and began establishing a base camp. For their part, the local officials greeted Stone, Vizcayno, and the commission with outward courtesy, although some thought the chief to be a bit of a blowhard. They informed him that the real authority in the state rested with Sonora's shrewd and charismatic governor, thirty-seven-year-old Ignacio Pesqueira, and that he would have to give his approval before the survey could proceed. A native of the rough frontier of northern Sonora, Pesqueira had been educated in Spain but returned to engage in campaigns against the Indians and political intrigue that marked the early years of the Mexican Republic. By 1856, he was a colonel in the National Guard, and in that year led a revolt against the sitting governor that ended with him in virtual control of the state with the titles of constitutional governor of Sonora and provisional governor of Sinaloa. By all accounts, Pesqueira was a shrewd, calculating military and political leader, highly protective of his state and its resources, as Stone and the commission soon would find out.[10]

The day after his arrival in Guaymas, Stone had what he believed to be a cordial informal meeting with Pesqueira, informing him of the survey's nature and mission. On April 15, he and Vizcayno formally presented the governor with the Mexico City orders and requested that instructions be issued to soldiers and civilian officials to assist the commission so that it could begin surveys in the Sonoran interior. While awaiting Pesqueira's reply, Stone received some disturbing news. Calhoun had declined to pay the drafts that he had drawn on him for commission expenses. Stone now had to rely on Jecker, Payno, Escandon, and Isham for funding, which would prove a more complicated process. The financial status of the commission, however, quickly took a back seat to other troubles when Pesqueira issued his reply. The governor apologized, but stated that, despite the orders from Mexico City, he lacked the authority to allow the survey to go forward. Permission would have to come from the state legislature. In addition, he had some questions regarding the orders that he would need to submit to the central government for clarification before he could offer assistance to the survey, and that could take quite a bit of time. All in all, Pesqueira's actions, and the subsequent legislative repudiation of the 1856 Jecker, Torre contract, amounted to a legalistic and polite, but firm, refusal to allow Stone and his men to chart the inland regions of Sonora. Less polite was an order issued by the governor demanding that Vizcayno leave the state—an order with which the judge prudently complied. He returned to Mazatlán on the next boat, asking Stone to send him any maps produced for his review there. From that safe distance he also continued to declare Pesqueira's decrees hindering the commission in its work and sending him home to be illegal.[11]

Stone was flabbergasted by Pesqueira's action. He believed orders from the Mexican central government would compel the governor to assist him, or, at least, not stand in the commission's way. In the United States, a directive from the Executive Department or Congress carried considerable weight, but what the survey chief did not realize was that on the Mexican frontier at the time the real power rested with governors or regional strongmen. As a consequence, Pesqueira might pay nominal allegiance to Mexico City, but he could do pretty much as he pleased with little worry about interference from the capital

The reasons behind Pesqueira's action are readily understandable. To begin with, the Sonora Survey Commission was a creature of Jecker, Torre & Co., already despised in the state because of its association with Raousett-Boulbon. This latest venture could very well be a cover for a similar scheme. Also, the commission originated in San Francisco, long a breeding ground for filibustering plots. Raousett-Boulbon had come from there, as had Henry Crabb, leader of an 1857 attempt to establish a colony in Sonora that had ended with his defeat at the Battle of Caborca and the deaths of Crabb and almost all of his followers, as well as the notorious William Walker, who had earlier in the 1850s led unsuccessful filibustering expeditions into Sonora and Baja California. In addition to the fear of invasion was concern on the part of many of Pesqueira's wealthy ranchero supporters, who had been freely grazing their herds for some time on the public domain, that they might lose this privilege if the *terrenos balidos* were surveyed and a substantial portion of the land turned over to Jecker or the Iturbide heirs. All of these factors combined with Stone's own tactlessness in dealing with local authorities to make the Sonoran government hostile toward the little surveying party.[12]

Stone was in a quandary. The special orders from Mexico City appeared not to be worth the paper they were written on, and the judge who was supposed to support him had turned tail and fled south at the first sign of opposition. As he saw it, he now had three choices. He could pack up the expedition and return to California, a failure again. If he did this, he also would likely be held liable for at least some of the expenses he had already incurred, putting him back where he had started—in debt and out of work. Or he could stay put in Guaymas and appeal to the Mexican government to use force to compel Pesqueira to reverse his position and assist the survey. Communications between Sonora and Mexico City, however, were notoriously slow, as was the decision-making process in the capital. Stone and his men might wait months for the central government to take action, if it even chose to do so, all the while losing valuable time to complete the survey within the terms of the contract. Or he could ignore Pesqueira and pursue the survey anyway, hoping that the Mexican government would come through with its support or that his employers might use their influence in Mexico City and Washington to pressure the Sonorans to abandon their opposition to the commission.

Stone chose the third option. It was a risky move, but one he felt he had to make. The survey was well armed, but probably could not prevail in a pitched battle against the governor and his troops. On the other hand, it was well known that Pesqueira had a number of political and military issues vying for his attention in the central portion of the state and in Sinaloa. So long as the survey parties stayed away from his power base in the capital of Hermosillo to the north and inland from Guaymas, they might just be able to accomplish at least a good part of their mission.

Throughout the summer and into the fall of 1858, therefore, the surveyors remained busy, locating and mapping hundreds of thousands of acres of public domain in western Sonora. Stone marveled at the agricultural richness of the Yaqui, Mayo, and Fuertes river valleys. In a booklet entitled "Notes on the State of Sonora," he later wrote:

> The foreigner will find himself wondering at the luxuriant crops produced by the imperfect cultivation in use there, and at the broad leagues of excellent lands left uncultivated for want of a little outlay of labor in clearing them and supplying them with the necessary irrigation.[13]

He was likewise impressed with the potential for development of silver, copper and lead mines. Summing up his evaluation, he wrote one of the investors, "You cannot . . . conceive of the value you have. You have the foundation of one hundred great companies in your contract—great land companies and great mining companies."[14] In the process of carrying out its surveys, the commission also undertook some scientific investigations, sending zoological and mineralogical specimens to the young Smithsonian Institution in Washington. Naturally, Pesqueira was less enthusiastic about the commission's work. From Hermosillo he continued to issue proclamations declaring the survey to be an illegal operation under Sonoran law and demanding that it cease its activities at once.[15]

In the meantime, while Stone explored and surveyed, and Pesqueira fumed, events in San Francisco would alter the commission's financial support and political position. On October 6, 1858, James Calhoun finally found purchasers for one half of his interest in the Jecker Contract. Former congressman and vocal expansionist Samuel Inge, and J. Mora Moss, one of the wealthiest men in California, bought in. Their entry promised to put the project on a more even fiscal footing and provide it with additional clout in the halls of government. Their presence undoubtedly gave Stone new confidence that he could prevail in the confrontation with Pesqueira.[16]

This hope, however, dimmed in the fall of 1858 when the official reply of the Mexican government to the protest he had filed back in April at last arrived in Guaymas. As he had anticipated, authorities in Mexico City upheld the protest. Yes, the bureaucrats stated, the commission had every legal right to continue its work and yes, Pesqueira ought to cooperate with Stone, but nowhere was there any mention of troops or other tangible actions to make it so. For his part, the governor proved just as stubborn as his adversary. He fired off a stinging reply to Mexico City maintaining that his actions had been in response to an imminent threat of annexation of all or part of Sonora by the United States, and that the commission was an agent for this aggression. In an outright act of defiance, he concluded, "In regard to the protest which you esteem convenient to direct to this government, I ought to say to you that this government does not consider it as just or legal."[17] Pesqueira understood that by issuing the order as written, the Mexican leaders were stating that this was as far as they were willing to go in support of the commission, and that they had no intention of forcing him to comply. Stone may have had the legal high ground, but it would do him little good if Pesqueira continued his opposition.

Pesqueira now felt he had free rein to turn up the heat on the commission. While affairs in Sinaloa still demanded his immediate personal attention, he instructed his subordinates to hinder its progress in every way short of provoking a direct confrontation. As the stalemate continued, the surveyors kept working, albeit plagued by obstreperous local officials emboldened by their chief's direction and Apache Indians raiding from Arizona into Sonora. To keep his employers and the United States government informed of affairs in Sonora and the commission's progress (or lack thereof), Stone used his lead engineer, Jasper Whiting, as a courier to carry messages to and from the Overland Mail line in southern Arizona. Occasionally, Whiting would transport confidential dispatches to San Francisco and Washington as well. Stone also did what he could to circumvent Pesqueira's roadblocks. He limited surveying largely to the less-settled parts of the state, and, in addition to the Guaymas base, established a camp for his engineers; one observer described it as "a little village," across the international line. From there, the surveyors could keep working in northern Sonora with a bit less harassment by Mexican authorities, and the presence of American dragoons at nearby Fort Buchanan offered some protection from the Apaches.[18]

Into this already volatile situation sailed an American sloop-of-war, the twenty-two-gun USS *St. Mary's*, a fixture of ten years with the Pacific Squadron. The ship had been ordered by the secretary of the navy to investigate the alleged mistreatment of some American citizens doing business on Mexico's West Coast. The source of this information is unclear, but it is quite possible that the new investors in the Jecker Contracts, Moss and Inge, had requested the navy investigate what was going on in Guaymas. The *St. Mary's* captain, Commander Charles Henry Davis, was a veteran of twenty-five years at sea. He had some experience in dealing with political matters in Latin America, having served as an officer on the sloop-of-war *Vincennes* and ship of the line *Independence* in Ecuadorian, Brazilian, and Argentinian waters. As captain of the *St. Mary's* he had pressured the Nicaraguans in 1857 to allow the surrender of filibusterer William Walker and his men, along with their safe passage out of the country. Upon arriving in Guaymas, Davis learned that the regular United States consul there, Robert Rose, had been away for some time and that no one really knew when he might return. He decided that because of the importance of the port in West Coast trade, such a state of affairs could not continue, and on December 16 appointed Stone as temporary consul. This move altered the survey chief's status immensely. Instead of just being the head of a private commission, he was now, at least in the eyes of the commander of the most powerful warship in the region, the official American representative in a foreign port. Almost at once, the animosity between Stone and Pesqueira, which had been pretty much a local dispute, began to have all the ingredients of an international incident.[19]

By this time, the pressures of managing the survey and sparring with the wily Pesqueira had started to tell on Stone. Usually even-tempered and genial, he now became irritable and even belligerent. A streak of anti-Mexican prejudice also came to the surface. Perhaps because he felt he had the cannon of the *St. Mary's* to back him up, Stone exceeded the usual consular duties of serving as an agent assisting American merchants doing business in Guaymas and sending reports to the State Department on conditions and events in the region. Instead, he let it be known that he would defend every American in the state against wrongs real or imagined. Any charges leveled against *yanquis* by Sonoran authorities must be false, and he would employ bluster and intimidation to get them reversed. Stone also fell in with those (including his backer, Inge) who proposed that the United States acquire all or part of Sonora and Baja California. Soon after assuming the consulship, he wrote to Secretary of State Lewis Cass that most "intelligent" residents of northern Mexico would welcome annexation. Such a course, he maintained, would be "the only means of saving this state [Sonora] from a return to almost barbarism."[20]

Pesqueira's biographer has also suggested that Stone and Commander Davis wanted to provoke an incident that would prod the United States to land troops at Guaymas and at least establish a protectorate over the region. While the two men likely would have approved of a protectorate, however, there is no evidence that they conspired to precipitate such an action. Davis seemed more concerned with carrying out his orders to check on alleged actions by Mexican authorities impacting American merchants than in launching an invasion, and Stone's behavior alone was enough to harden local sentiment against him, the commission, and the United States.[21]

The first serious incident came about a month after Stone's appointment. An American citizen in Guaymas fired his pistol at a stray dog whose barking was annoying him. The bullet missed the offending dog, ricocheted, and struck a woman and small child, injuring both slightly. Although he quickly apologized and offered to pay all medical expenses, Guaymas police charged the shooter with violating a town ordinance against discharging weapons and took him to jail until the matter could be sorted out. Rather than calmly negotiating the prisoner's release, upon learning of the arrest, Stone stormed to the jail arrogantly demanding that the American be freed immediately and hinting of dire consequences to the town from the warship in the harbor if he was not. The American was soon let go, but from a diplomatic standpoint, Stone's actions accomplished little more than to antagonize the Mexicans.[22]

A couple of weeks later, an American trader at Guaymas named Goerlitz reported some weapons stolen from his warehouse. The guns were supposed to be sold to the Sonoran military, but when the buyer could not pay cash Goerlitz refused his draft and called off the deal. Accusing the merchant of bad faith, agents of Governor Pesqueira then entered the warehouse, took the weapons, and left in a boat. Once again, Stone sprang into action, gathering the captain and crew and setting sail in the *Manuel Payno* in an unsuccessful search for the stolen goods. He returned empty handed, but the subsequent arrival at Guaymas of another American sloop-of-war, the USS *Vandalia*, induced the prefect of the town to make an amicable settlement with Goerlitz. The local citizenry, nonetheless, regarded Stone's actions as unwarranted and not a little foolish.[23]

In addition to his bullheaded diplomacy, some of Stone's actions as chief of the commission angered and worried Guaymas residents. When the surveyors were not in the field, Stone armed and publicly drilled them in a military fashion, even going so far as to march his staff up and down the town's principal streets. While such actions may have been meant as a warning to Pesqueira not to interfere with the survey, it stirred memories in many Sonorans of

Raousett-Boulbon, Crabb, and Walker. Stone's conduct definitely gave the heretofore civilian commission the appearance of a filibustering party. In addition, when Commander Davis needed to send a detachment of sailors to "protect American interests" in Mazatlán, Stone obligingly provided him with the *Manuel Payno* and her crew. A navy lieutenant from the *St. Mary's* took command of the brig, which Davis immediately commissioned as an American ship of war, and with ten armed sailors set out for Sinaloa. Although Stone had the authority to do what he wished with the *Manuel Payno*, the Mexicans viewed his act as aggressive and worried about how quickly she had been transformed from a private survey vessel into a warship.[24]

Pesqueira, in the meantime, remained concerned with matters in Sinaloa and did little more than issue a protest of Stone's appointment and maintain that he would refuse to recognize him because he lacked a formal commission from the State Department and an exequatur from Mexico City.[25] The return to Guaymas of Robert Rose in January 1859 further muddied the waters, since Stone did not immediately relinquish his consulship. Rose also lacked an exequatur, so the Sonorans ignored him as well. The result was two consuls, neither of whom the Sonorans liked or recognized. By late February, it had become obvious even to the Americans that two consuls at Guaymas were too many, so when the steam sloop-of-war USS *Saranac* arrived in the harbor, Stone submitted his resignation to her captain. Stone's exit from the consulship, however, did little to ease the brewing conflict. Rose favored a hard line with Pesqueira in matters involving American citizens, including the Sonora Survey Commission, and he did not hesitate to call on the navy to back up his position. He and Stone still cooperated in stirring the pot, sending a regular flow of complaints to Washington about alleged insults to the United States and urging the permanent stationing of a warship at Guaymas to protect Americans travelling and doing business there. The situation came to a head when word reached Guaymas of President Buchannan's second annual message to Congress, in which he mentioned lawlessness along the border and asked for a temporary protectorate over Sonora and Chihuahua. Perhaps Stone's reports of dissatisfaction among northern Mexicans with their government had reached the president, for in his message he indicated a full-scale invasion would not be necessary because the people would welcome de facto annexation by their neighbor to the north. He followed up this request with a suggestion that the United States buy portions of Sonora, Chihuahua, and Baja California outright. Although Congress took no action on the president's proposals, the mere hint of a protectorate or purchase infuriated the Sonorans. For some time they had suspected Washington of secretly supporting filibustering attempts directed against Mexico, and now the Yankee leader had publicly stated that he wished to acquire a sizeable portion of their state. Agitation at Guaymas against Americans in general and Stone in particular increased substantially over the next several weeks, causing Rose to redouble his efforts to get a warship stationed in the Gulf of California, and giving Pesqueira the opportunity he had been waiting for to rid himself of the pesky surveyors. Responding to his peoples' patriotic fervor, on May 18, 1859, he issued a proclamation officially ordering the commission to leave Sonora within forty days.[26]

Although they must have seen it coming, Stone and Rose seemed surprised, and immediately sent protests against the expulsion to Hermosillo, Mexico City, and Washington. Sentiment among the people of Sonora, however, made it clear that Pesqueira had the state solidly behind him. About this time he also returned from Sinaloa with a formidable military force. Remaining past the forty-day deadline without intervention by the navy could prove hazardous to the commission, and the *St. Mary's* had left Guaymas for Panama, leaving no United States warships in the vicinity. Besides, Stone's support among the merchant community on Mexico's West Coast had begun to erode. Too much anti-American agitation was bad for business. The consul at Mazatlán even went so far as to write the State Department that Stone's continued presence and belligerent attitude could lead to serious consequences for all Americans in the region.[27]

Stone had no choice but to evacuate, but not back to San Francisco. He sent the *Manuel Payno* to Baja California, ostensibly to establish a base for staff who would later work on the Iturbide Grant, while he led the main body of surveyors from Guaymas north across the border to the site he had established a year earlier in the Sonoita Valley southeast of Tucson, about twenty miles from Fort Buchanan. They named the heretofore anonymous location, appropriately, "Camp Jecker," and from there they would be "ready to return and resume the . . . survey and exploration," once sufficient diplomatic and/or military pressure could be brought to bear on Pesqueira to relent.[28] Until then, life at the camp settled into a routine of warding off raiding Apaches, as well as prospecting in the nearby Santa Rita Mountains and paying occasional visits to Fort Buchannan and Tucson. Stone, meanwhile, boarded a stagecoach for Washington to press the commission's case. At the end of August, he addressed a passionate memorial to the secretary of state vigorously protesting the expulsion and pleading for him to take any steps "deemed proper and necessary for the maintenance of treaty stipulations and the safety of the rights of your petitioner and his comrades."[29] Jecker, Moss, Isham, and Inge were also working behind the scenes in Washington and Mexico City to get the commission reinstated and their contract back on track. By the fall of 1859, all of these efforts seemed to bear fruit. The United States government was ready to take action to resolve the dispute, but a lack of communication among the parties in Arizona and Sonora would lead to an even more serious confrontation.

On October 5, 1859, the *St. Mary's* sailed back into Guaymas harbor with a new captain and crew. If the Mexicans thought that this change marked a more conciliatory approach by the Americans, they soon recognized their mistake. Her skipper, Commander William D. Porter, a cousin of Stone's West Point classmate and Mexican War comrade Fitz-John Porter, had orders to compel Pesqueira to readmit the commission. Upon arriving, he sent the governor a letter lecturing him on the seriousness of the situation by stating that Washington would not have sent the *St. Mary's* on a voyage of some two thousand miles for no good reason, and threating severe action if the expulsion remained in effect. Porter further insulted Pesqueira by strongly hinting that Sonora would be better off under an American protectorate than under his administration. For the time being, Pesqueira ignored him. Porter, nevertheless, continued to bluster and threaten. The guns of the *St. Mary's* could clearly inflict severe damage on the port city, but the Jecker Contract and the Sonora Survey Commission had now become matters of principle and national honor. The Mexican authorities, therefore, politely but firmly refused to acquiesce.[30]

At the end of October, Porter issued an ultimatum that

Pesqueira meet with him face to face to resolve the Stone issue. The governor agreed, but to back up his bargaining position, Porter landed a force of one hundred sailors and marines, along with two cannon, and deployed them a short distance from the meeting place. For his part, Pesqueira brought eighty-five cavalrymen. Both sides began by calmly stating their arguments, but when Pesqueira flatly refused to reverse Stone's expulsion, Porter broke off the negotiation and returned to his ship, threatening a blockade. This alarmed the merchants of Guaymas, both foreign and Mexican. The town's economy depended on trade, and consuls from France and Spain met with Porter and urged him to take a softer line. The commander agreed not to blockade, but refused to budge on his demand to readmit the commission. He even upped the ante by asserting the right to fly the American flag above the consulate on national holidays of the United States and Mexico. By international agreement, the Americans had this right, and the Sonoran authorities quickly agreed to Porter's terms regarding the flag. To the residents of Guaymas, however, the arrogance of Porter's demand was the last straw. The unrest bubbled over with an invasion of the consulate on November 18 by a gang of soldiers. The unruly mob tore down and trampled the American flag and consular seal. A few days later, another mob assaulted the vice consul, Farrelly Alden, who had been appointed earlier in the year by Rose prior to leaving his post, and who was officially recognized by both Pesqueira and Porter. At this point, Pesqueira decided that things had gone far enough. He certainly did not want the situation spinning further out of control and resulting in a war with the United States, which he knew he could not win. Through the prefect of Guaymas, he issued an apology for the incident and ordered the citizens and soldiers to leave the consul alone. The governor also personally took the desecrated flag to the consulate and presented it to Alden. Under normal circumstances this would have settled the matter, but Porter's blood was up. He did not think the apology abject enough, and stated he would bombard the port. The Americans and other foreigners living in Guaymas, frightened by what could happen next, intervened and managed to calm the commander down. He stayed the attack and through back channels reopened negotiations with Pesqueira. It finally appeared as though cooler heads might prevail.[31]

While Guaymas settled into an uneasy calm, events were transpiring in Washington that got the pot stirring again. Stone, upset with the navy's seeming inability to effect the commission's return to Sonora, had decided that his old service branch could produce better results. He still had some connections in the War Department, and wrote Secretary John B. Floyd asking that an officer be sent to Guaymas from Fort Buchanan to secure repeal of the expulsion order. Floyd agreed, and in early November dispatched Captain Richard S. Ewell of the Second Dragoons to Sonora with orders from the adjutant general to issue a formal protest against Pesqueira's actions. Rather than a mission to coerce the Mexicans, however, the adjutant general referred to the assignment as a "delicate and responsible duty." He would be more of a mediator than a bully. To the Mexicans, Ewell's presence nonetheless represented an unwanted intrusion into their affairs by a military representative of an increasingly hostile power. He knew he would not get a particularly friendly reception.[32]

When he finally arrived in Guaymas, Ewell expected to walk into a confrontation, but he found that Porter and Pesqueira had "amicably" adjusted the Stone dispute. The commander told Ewell that he understood the surveyors would be allowed to return and continue their work. He suggested that the captain let the matter drop, something that he was only too happy to do. After a courtesy call on Pesqueira, Ewell was on his way back to Fort Buchanan. While passing through Hermosillo, however, he was arrested on allegations of stealing a mule. He denied the charge and refused an offer of the alleged owner to loan him the animal. Ewell even suggested he might pay for the mule so that he could get on with his journey, but the local judge saw things differently. He confiscated the mule and held Ewell under a substantial bail. Ewell ignored the judge and decamped to Guaymas, where he complained to Porter about the incident. Any understanding between Porter and Pesqueira quickly evaporated. The commander once again rolled out the guns of the *St. Mary's* and grumbled about bombarding the port. This time the Mexicans decided to comply, and released Ewell, although his possessions remained in custody. He also was given an "escort" for the uneventful ride to the border. In subsequent talks, Porter was able to retrieve Ewell's baggage. The *St. Mary's* then sailed from Guaymas, her captain, in all likelihood, believing the Stone/Ewell affairs to be settled.[33]

Throughout the winter of 1859–1860, Pesqueira delayed readmitting the commission. Once the *St. Mary's* was safely far away, he clarified his position, which was exactly the opposite of what Porter felt he had negotiated. Pesqueira stated that he indeed had agreed to readmit Stone and the surveyors, but only as individual private citizens and not as part of an expedition. They could not engage in any organized surveying under the Jecker, Iturbide or any other contracts. From the Mexican standpoint, this carefully worded decree effectively ended the work of the Sonora Survey Commission. Many of the commission staff, nonetheless, remained in southern Arizona for a time working on a government wagon road improvement contract that Stone had secured, hoping to keep them around while he continued to request the *St. Mary's* return to Guaymas and change Pesqueira's mind. When the road contract was up and their salaries stopped coming, however, they abandoned Camp Jecker and left the Southwest. The *Manuel Payno* continued to sail in Mexican waters, taking whatever cargoes she could get until running aground in March 1860.[34]

Stone relocated to Washington, where he and his family moved in with his in-laws, the Clarys. He devoted himself full-time to lobbying on behalf of the Jecker contract, and in February 1860 issued a tract in English and Spanish "To Whom It May Concern" ("*A todos a quienes intersare*"), in which he recounted the history and legality of the Jecker contract, as well as the abuses the commission had suffered at the hands of the Sonorans. The publication ended with a warning to potential purchasers of surveyed *terrenos balidos* in Sonora not to enter into any agreements without first obtaining permission from J. B. Jecker & Co., the legitimate owner of one-third of those lands. For their part, Jecker and Isham also pushed the American and Mexican governments for stronger action on behalf of the commission, and sought to recoup money they claimed to have expended on the survey. Their cause received a boost in February 1860, when the distinguished American statesman and jurist Caleb Cushing issued a judgment that the 1856 contract was:

> Valid in form and substance; that it was a good contract for the Mexican Government to make; that it is the right of the contrac-

tors to complete the survey according to the contract; and that the contracting parties are now entitled in law to hold, and do hold in full property, the undivided third part of all the lands surveyed by them under the contract in the State of Sonora.[35]

Senor Monteverde, secretary for the state of Sonora, soon answered Cushing's judgment in a lengthy paper, which refuted point by point, at least from the Mexican perspective, the American jurist's assertions. In the end, despite aggressive lobbying and Cushing's opinion, nothing official was done on behalf of Stone, Jecker, and Isham. The tide of diplomacy and public opinion had turned against them. The United States had reestablished formal diplomatic relations with Mexico, suspended since the summer of 1858, and was negotiating with the government of Benito Juarez for a treaty granting transit and commerce rights across the Isthmus of Tehuantepec and northern Mexico, as well as the right to intervene militarily if these routes were threatened. The State Department did not want a formal protest over the expulsion of the survey to upset these talks. Besides, by 1860 the Mexicans had finally cancelled the Jecker contract, and pressing the matter further could weaken the position of the liberal Juarez, whose Republican faction the Buchanan administration supported in the "War of Reform" against Mexico's conservatives. By early 1860, therefore, Washington was content to let what the *New York Times* had called "a Western War Cloud" dissipate. The paper's editors reflected the sentiments of the government and many private citizens when they wrote that the Jecker contract matter was a Mexican domestic affair, and that the United States had done enough on behalf of Stone and the commission: "It is difficult to conceive on what ground that firm [Jecker & Co.] . . . can ask our troops and ships of war to move upon Sonora and enforce a contract declared by that State to be illegal." Furthermore, the interference of Commander Porter was seen as "a blow struck against the Juarez, or Constitutional Party of Mexico on the Pacific."[36] Nor was the *Times* alone in this belief. When Porter issued one of his threats to shell Guaymas, San Francisco's *Evening Bulletin* had warned, "Capt. Porter . . . may bombard Guaymas; but if he does so the act will assuredly be disavowed by the President, or the latter will be called to an account by Congress." The paper went on to remind its readers that "it has never been the policy of the States to seek to enforce fulfilment of contracts with its citizens . . . with foreign governments. On the contrary, it has always peremptorily refused to become sponsors or guarantors of such contracts."[37]

With government and public support for the commission waning, the survey, or what was left of it, was officially withdrawn from the Southwest in the fall of 1860. The Jecker and Isham claims themselves did not die as quick a death. In Mexico City, Jecker kept protesting the commission's expulsion and seeking indemnification for his alleged loss in Sonora, but to no avail. This claim, however, was minor in comparison with other demands he placed on the Mexican government. Earlier he had loaned Mexico 3.75 million French francs. As security he had received Mexican bonds, which, in 1861, he insisted be redeemed at a face value of 75 million francs. This amount, along with the Sonoran property and other debts that Jecker, still a French citizen, purported to be owed him by Mexico, were among the grievances cited by Napoleon III to justify his invasion of that country the following year. The French intervention did not work out well for Jecker, and after the downfall of the Emperor Maximilian, he and Isham sold a majority of their interest in the Iturbide Grant and the 1856 contract, by now a dubious document, to say the least, to the Lower California Company. This New York corporation then filed a $12.5 million claim against Mexico for lands mapped in Sonora, Sinaloa, and Baja California by the Sonora Survey Commission and the Iturbide heirs, and called for the right to a complete survey of Sonora's *terrenos balidos* begun by Stone. At this time, the Lower California Company was also engaged in a colonization scheme in Baja California. When the venture proved to be fraudulent, however, the Mexican government cancelled all of the company's contracts and concessions. The firm's investors attempted to persuade the United States to seek an indemnity against Mexico for their alleged loss, but in 1872 the State Department denied their request. Although the claim of the Lower California Company would remain in the courts for some time, Washington's action ended for good the matter of the Jecker contract and all possibility that the work of the commission would be resumed.[38]

The commission's nemesis, Ignacio Pesqueira, continued to govern Sonora off and on into the 1870s. He and his family removed to Arizona when the French invaded in 1865, but he returned the following year to lead the Republican reconquest of the state. Over the next decade, Pesqueira attempted to bring some stability and economic development to Sonora, although this was made difficult by military uprisings, political turmoil, and Apache depredations. By the mid-1870s, he had lost the support of a majority of Sonorans, and in 1877 he retired from public life for the last time to his ranch, *Las Delicias*. Pesqueira lived there quietly and pursued various business interests until his death in January 1886.[39]

As for Stone, he had little else to do but remain in Washington, dutifully completing maps of the hundreds of thousands of acres the commission had managed to survey and publishing "Notes on the State of Sonora." But few Americans took an interest in this description of the possible wealth in the lands of their southern neighbor. Their own country seemed on the verge of coming apart as the debate over slavery and its extension into the western territories intensified. A critical presidential election was in the offing, and talk of secession was spreading throughout the Southern states. As he read newspaper accounts and listened to Washington gossip about the crisis Stone must have speculated on how it might affect him and his family. The events of the coming months would indeed impact them, but in ways beyond anything he could have imagined.

DEFENDING
THE CAPITAL

As the fall of 1860 turned into bleak winter, it was clear that the Sonora Survey Commission had reached a premature end. The cost of the project was unclear and no additional funds for its completion would be forthcoming. Juan Baptiste Jecker claimed to have spent over a million dollars on his contract. Other estimates were as low as $100,000 to $250,000. But no matter the cost, none of Sonora's *terrenos balidos* mapped by the survey would be transferred to the investors, and none of the participants would realize the wealth the Jecker Contract had promised. The commission's chief, and last employee, Charles P. Stone, remained in Washington, completing his reports as well as the maps and charts his engineers had plotted. He and his family still lived with his in-laws, Captain Robert Clary and his wife, in their First Ward home on the city's northwest side. The arrangement must have been uncomfortable for the thirty-six-year-old Stone. When he had wed their daughter, Maria Louisa Clary, seven years earlier he had been an up-and-coming young army man with a highly responsible position as ordnance officer for the Pacific Coast. Then came his embarrassing failure as a banker, not to mention the loss of Captain

Clary's money, and the collapse of the Sonora venture. Now he was dependent on the Clarys for the roof over his family's head.

Stone's work for the commission would soon come to an end, and he likely sought other employment, but, like most everyone else in Washington, employers were more concerned with events transpiring to the South than with adding to their payrolls. The election of Abraham Lincoln as president had stirred rumblings of secession in the slaveholding states. On December 20, South Carolina became the first state to leave the Union. Others seemed sure to follow, compounding the crisis. Officially, the administration of President James Buchanan did little other than protest the act, leaving the matter to simmer until his successor could be inaugurated.

A perceptive and staunchly Union man, Stone was keenly aware of what was going on in the District of Columbia and beyond. The nation's capital occupied a unique position. Locally administered by a mayor and aldermen, its purse strings were controlled by Congress, which was itself divided between North and South. The selling of slaves in the district to other parts of the country had been banned by the Compromise of 1850, but slavery itself still flourished among the many households of southern politicians, bureaucrats, soldiers, and businessmen. Their presence gave the city a decidedly Southern air and worried those wishing to remain loyal to the Union. To make matters worse, the district was surrounded by Virginia and Maryland, slave states whose allegiance was questionable. Stone's appreciation for Washington's situation and potential peril would prove crucial when, perhaps looking for a job, he paid a visit to his old commander, General Winfield Scott.

At age seventy-four and a veteran of every American conflict since the War of 1812, Scott still commanded the United States Army. He had been "exiled" for political reasons some years earlier, but Buchanan recalled him to Washington amid the growing secession crisis. The transfer of his command back to the district had a calming effect on the Unionists. "God bless you, General!" some would shout as his carriage navigated the muddy streets between his lodgings at Wormley's Hotel on I Street and his headquarters in the Winder's Building on Seventeenth Street opposite the War Department. Of course, not all Washingtonians welcomed his presence. Threats were made against him, which he ignored.

He had more serious matters on his mind—a growing rebellion to handle and a capital to defend.[1]

The army Scott commanded numbered just over sixteen thousand officers and men, the bulk of whom were scattered at outposts in the western states and territories and at isolated coastal forts. Many in the officer corps were Southerners, whom Scott, a Virginian, had favored in promotions and assignments. He liked their military bearing, good humor, and pleasing manners. A number of these had already resigned their commissions to serve their native states, and more would leave the army as their states seceded. On the other hand, many promising officers from the North, such as Stone, Sherman, and McClellan, saw few opportunities for advancement and left the army in the 1850s. Looking around Washington itself, Scott found only a few hundred marines, under the jurisdiction of the Navy Department at the Navy Yard, an ordnance company at the arsenal, and a handful of regulars and clerks at the War Department. There were militia organizations such as the National Rifles and Potomac Light Infantry, but the general knew almost nothing about their training, ability, or loyalty. Besides, he was a regular army man who did not think much of the competence of volunteer soldiers, loyal or disloyal. Scott's knowledge of the district itself also was out of date. For the past five years he had maintained his headquarters in New York City and resided in a comfortable brownstone that admirers had purchased for him there. He had taken to the field on occasion, including a taxing expedition to the Pacific Northwest to resolve the "Pig War" that threatened conflict between the United States and Great Britain. In addition, Scott's age, associated physical infirmities, and workload would prevent him from getting out and effectively gathering the intelligence needed to suppress sedition and mount an effective defense of the capital. He needed a trustworthy and energetic subordinate to be his eyes, ears, and legs, and on New Year's Eve of 1860, just such a man came knocking on his door.[2]

On the chilly evening of December 31, 1860, Stone paid a call on Winfield Scott at his Wormley Hotel residence. He was likely surprised at the general's appearance. During the Mexican War, Scott had been an energetic man in his early sixties, but the ensuing twelve years had not been kind. Now he was overweight, jowly, and suffering from rheumatism, dropsy, and a host of other ailments. He had difficulty standing and could barely walk. Riding was out of the question. Once the two began to converse, however, it became obvious that Scott's faculties and military acumen were little diminished. For a while they reminisced about old times in Mexico, each man sizing up the other. How much Scott remembered about Stone is difficult to surmise. He had been present at Stone's military academy examinations, but that was long ago. He had also commended and brevetted the young ordnance lieutenant during the Mexican War, but such honors he had bestowed on many other junior officers. Perhaps he knew of his tenure as ordnance officer for the West Coast or had followed Stone's recent ups and downs in Sonora in the pages of the *New York Times*. Nevertheless, Scott seemed in an upbeat mood, and referred to Stone as his "young friend." Then Stone brought up the state of the nation and its capital, and the conversation took on a tone of urgency. Scott remarked that neither conciliation nor force alone would resolve the secession crisis. He favored a combination of the two policies in dealing with the South, and intended to discuss that with President Buchanan when he met with him in about a quarter hour. After some slow pacing up and down, he turned to Stone and asked, "How is the feeling in the District of Columbia? What proportion of the population would sustain the Government by force, if necessary?" Having already studied the situation in the capital and inquired of some of its leading citizens about loyalty in the event of an insurrection, Stone readily replied that two-thirds of the district's "fighting stock" would be true, but that they were disorganized and without a "rallying-point." As the two men approached the general's waiting carriage, Scott turned to Stone, placed his hand on his shoulder, and said: "Make yourself that rallying-point."[3]

Why did Scott suddenly choose Stone to be his right-hand man? After all, the army contained more experienced officers, including the brilliant Robert E. Lee, whom Scott had marked for a high command, maybe his own. But Scott needed someone right away, and most officers, including Lee, were on duty far from the capital. Perhaps no one in Washington had yet impressed him. Or, perhaps, he did not trust the loyalty of those who did impress him. The capital had a number of staff officers on hand, but some of these were near, at, or beyond the retirement age, while others had not held field commands for some time. They lacked, as one observer noted, "dash." Also among the bureau chiefs and other military bureaucrats were Southerners, whom it was suspected might defect to the seceded states—and a number did. No doubt, Scott saw Stone as the best man available at the time for the job of protecting Washington from internal sedition as well as threats from the surrounding slaveholding region. For his part, Stone demonstrated the energy, enthusiasm for the Union, capacity for leadership, and knowledge of conditions in the district badly needed by Scott. In addition, Stone was a West Pointer—an army regular on whom the old general felt he could always depend—with some combat experience in Mexico and a reputation as an able and creative administrator during his time at Benicia. While Stone probably had visited Scott with the idea of offering his services in some capacity to the general, he had not anticipated receiving such an order. One can imagine him rushing home to tell Maria, Hettie, and the Clarys that he had just been tasked with defending the national capital![4]

The next morning Stone presented himself at the White House to receive his commission as a colonel and the inspector general of the District of Columbia militia. On January 2, he was officially mustered into the army, becoming, he would later claim, the first of an eventual half-million-plus men to serve in defense of the Union over the next four years. Stone was given the position of chief of staff for General Roger Weightman, the nominal commander of the District of Columbia militia, but in reality he reported to Scott. He was assigned an office in the War Department near that of the secretary and immediately began sorting out friend from possible foe among the district's four active militia units. As inspector general, Stone was able to utilize detectives to investigate the militia. He quickly learned that three companies, the Potomac Light Infantry, Washington Light Infantry, and National Guard, were, by and large, loyal, while two others—the National Rifles and a newly formed unit called the National Volunteers—were highly suspect. The National Volunteers were led by Doctor Boyle, a prominent and politically well-connected physician. The unit was said to consist of three hundred men, and was drilling nightly in a large hall over a livery stable. Even though the organization had not yet received its weapons, Stone knew better than to take it on directly. He settled on a subtle approach, assigning a detective to infiltrate and furnish him with information on any treasonable activities. Stone soon had ample evidence of the unit's pro-secession nature. When Boyle came to the War Department to deposit with the inspector general the muster roll, certification of election of officers, and other documents for a National Volunteers company that were required for the distribution of arms, Stone thanked him, locked the papers in his desk, and sent him on his way. Only then did the doctor realize that he had been duped into turning over a roster of disloyal men, with his own name at the top, to the authorities. He was soon reported to have left Washington, headed south. Without its leader and possessing just a few sporting rifles, the National Volunteers disbanded.[5]

Dealing with the National Rifles would be another, more complicated, matter. Composed largely of Marylanders whose leader, an Interior Department employee named F. B. Schaeffer, made it clear they would oppose any movement of federal troops across their native state, the company was organized, disciplined, and armed. On the second of January, Stone met with its captain and cautioned him against making inflammatory statements regarding what his company might do if provoked. He could tell right away that this man was trouble. Stone then launched an investigation of the National Rifles, and what his detectives found both surprised and worried him. In addition to a full complement of one hundred men (with more joining daily), all armed with muskets and possessing a substantial supply of powder and ammunition, the National Rifles had acquired two mountain howitzers with harnesses and gun carriages, and an extra stock of pistols and sabers. Normally, such infantry units would be authorized only muskets. Stone quickly set about to determine how the National Rifles had obtained such a formidable arsenal. The chief of ordnance told him that he had been ordered by the former secretary of war, Southerner John B. Floyd, to provide Schaeffer with whatever arms he requested. He also discovered that Floyd had recommended Schaeffer's promotion to major, which would have given him authority over other militia units in addition to his own. Stone returned to his office, marched down the hall and presented his evidence to the new secretary, Joseph Holt. Aghast at the actions of his predecessor, Holt issued orders that prohibited the issuance of any arms to militia units without Stone's approval and requiring that Stone personally deliver all commissions of militia officers in the district.[6]

With Secretary Holt's orders in his pocket, Stone set in motion a plan to trick Schaeffer and eliminate the threat posed by the National Rifles. When the militia captain learned that he could no longer requisition weapons without Stone's approval, he visited the inspector general and boasted that he would take what he wanted from the Washington Arsenal by force. Stone calmly informed him that if he did so, the National Rifles would be fired upon by federal troops. This was a bit of a bluff, since Stone had only one regular army artillery battery and some loyal militia at his disposal. A company of regular sappers and miners requested from West Point had not yet arrived. But the bluff worked. Schaeffer turned and left for his office to ponder his next move. Stone now went on the offensive, sending Schaeffer an order for the National Rifles to return the howitzers, pistols, sabers, and other equipment beyond that required by an infantry company to the district's Columbian Armory that day. The captain, still under the belief that Stone had sufficient force to compel him to do so, complied. He then returned to the War Department to inquire about his major's commission. Stone appeared to be very busy at his desk, but stopped his work and from a drawer took the commission along with an oath of allegiance to the United States. He handed the oath to Schaeffer and told him to sign it in the presence of a justice of the peace. He knew that the captain was, if nothing else, a man of honor, and given his pro-Southern sympathies he could not sign the oath. After a long pause and a bit of inconclusive banter back and forth, Stone declared that it appeared as though Schaeffer was declining his commission and returned the document to the drawer. Schaeffer denied that this was his intention and demanded that he be given the commission. Stone replied:

> But, sir, you cannot have it. Do you suppose that in these times . . . I would think delivering a commission of a field officer to a man who hesitates about taking the oath of office? Do you think that the Government of the United States is stupid enough to allow a man to march armed men about the Federal District under its authority, when that man hesitates to take the simple oath of office? No, sir, you cannot have this commission; and

more than that, I now inform you that you hold no office in the District of Columbia volunteers.[7]

When Schaeffer responded that he did, indeed, hold a captain's commission, and that he had accepted it in writing, Stone played his trump card. Regulations required that acceptance of a commission must be accompanied by an oath of allegiance. Since he could not locate such an oath, Stone shrugged, "You have never legally accepted your commission, and now it is too late. The oath of a man who hesitates to take it will not now be accepted."[8] Faced with a bureaucratic brick wall and realizing he had been outfoxed, a furious Schaeffer stormed out of Stone's office. He resigned both from the Interior Department and the National Rifles and departed the district, taking several pro-secession sympathizers with him. Stone quickly replaced them with loyal volunteers.[9]

It had been a busy week for the new inspector general, uncovering and diffusing two potential threats to the peace and security of the capital through subterfuge and red tape. But Stone's work was just beginning. The allegiance of the existing militia was assured, but Stone knew he would need to organize additional volunteer units to defend Washington against a possible invasion from the slaveholding states, and to ensure an uneventful transfer of power from Buchanan to Lincoln. To do this, he wrote forty of the district's foremost loyal citizens asking them to form one volunteer company each. Some declined or did not respond at all, but within six weeks enough had complied to form and begin drilling fourteen companies.[10]

Recruiting the volunteers was one thing; arming them proved to be quite a different matter. Once the companies had been given a sufficient amount of unarmed drill, Stone authorized their members to be issued weapons. Much to his surprise, however, these requests began coming back denied by the Ordnance Department. When questioned about the denials, the chief of ordnance replied that the prohibition had come from President Buchanan himself. Stone then dropped by Secretary of War Holt's office and asked why the president had given such an order, stating that unless he could arm the loyal militia, he may as well resign. The secretary agreed that it made little sense, and directed Stone to take his case personally to the White House. He found Buchanan sitting at a writing table, still in his dressing gown. In the last days of his administration, and wrestling with the secession crisis, the president appeared tired and careworn. Stone wasted little time with pleasantries and inquired directly about his refusal to arm the militia. The president responded that he had been advised to do so by Washington district attorney Robert Ould. A Southerner, Ould did not want to see a strong military presence in the district, and used the flimsy pretext that providing weapons to militiamen from the Northern Liberty Fire Company and its archrival the Lafayette Hose Company could lead to the two volunteer organizations shooting at each other rather than at insurrectionists or invaders. Buchanan had consented and issued the order in question to apply to all militia companies. Stone politely refuted Ould's argument, and when the president asked him if he would bear the responsibility for any disorder if the firemen were to receive their muskets, he readily agreed. Soon all the militia units were mastering the manual of arms and preparing to safeguard the capital.[11]

Abraham Lincoln was elected the sixteenth president of the United States in November 1860, but would not be inaugurated until March 4 of the following year. In the interim, he decided it prudent to determine the loyalty of the general-in-chief and learn as much as he could about conditions in the capital. Ever since his election, rumors had been circulating that Scott, a Virginian, might side with his native state if it, too, seceded, and that Washington itself was a hotbed of secessionist plotting. In January 1861, therefore, Lincoln dispatched the adjutant general of Illinois, and a trusted confidant, Thomas Mather, to meet with Scott to gauge his political leanings. Mather found the old general be a staunch Unionist and ready to defend the national government and its president to the last. Lincoln also sent Leonard Swett, another close associate who had practiced law with him on the Eighth Circuit in Illinois and who had played a key role in securing his presidential nomination, to Washington to assess the city's stability and overall military situation. Swett met with Scott, who referred him to Stone. He called on the colonel at home and introduced himself and his mission. The two men spent the next several days discussing conditions in the capital, as well as reviewing the militia and touring the Columbian Armory and Washington Arsenal. Before taking his leave, Swett expressed his appreciation to Stone for the, "vast amount of careful work," he and Scott had done, "to ensure the existence of the Government and to render certain the inauguration of Mr. Lincoln. He will be very grateful to you both." At that point, rather than graciously accepting the thanks, Stone replied, as he would admit later, "with more sincerity than tact":

> Mr. Lincoln has no cause to be grateful to me. I was opposed to his election, and believed in advance that it would bring on what is evidently coming, a fearful war. The work which I have done has not been done for him, and he need feel under no obligations to me. I have done my best toward saving the Government of this country and to insure the regular inauguration of the constitutionally elected President on the 4th of next month.[12]

It is not known whether Swett relayed Stone's remark to Lincoln, but what the colonel had meant as a profession of fealty to the Union would become public knowledge and later come back to haunt him.

Assured of the commanding general's loyalty and that the capital was not on the verge of descending into anarchy, Lincoln could go ahead with plans to journey east to the inauguration. In the meantime, the result of the election would have to be ratified by the Electoral College, which met in the Capitol on February 13, 1861. Scott fretted that its proceedings would be disrupted by secessionists or other opponents of the president-elect, and let it be known that any such action would be quickly and harshly suppressed. The event, however, passed without incident, thanks in great measure to Stone's work in gathering intelligence on potential traitors and weeding out the district's militia companies.[13]

Of far greater concern to both Scott and Stone would be the safe passage of the president-elect from Illinois to Washington, and the security of his inauguration. For some time, they had been receiving messages warning that Lincoln would be assassinated on or before March 4. Most of these they discounted as the work of crackpots, but more than a few came from reliable sources and required investigation. Although unable to uncover any definitive threat, they determined that the most likely location for an attempt on the president-elect's life as he travelled to the capital would be in Maryland. The state was decidedly pro-slavery, and the train carrying Lincoln and his party would be susceptible to ambush at a number of points along the tracks. In addition, there was no through line between Philadelphia and Washington. In Baltimore, the center of secessionist activity, they would have to navigate between train stations. An attack by a mob of "plug-uglies" or a lone gunman was a distinct possibility there.[14]

While travelling from Illinois, Lincoln also began receiving word of threats on his life. Detective Allan Pinkerton, who had been hired by the Philadelphia, Wilmington, and Baltimore Railroad to investigate possible sabotage along the presidential route, passed these warnings on to the president-elect and his associates. Despite concern among his entourage, Lincoln dismissed such information as vague and lacking specifics, and insisted on keeping to his original itinerary of stops and speeches as he proceeded east. The danger became clearer on the morning of February 21, when a New York City detective named David Bookstaver visited Stone. He had been sent to Baltimore by the New York chief of police to look into rumors of a plot to kill Lincoln. What he revealed prompted Stone to take immediate action. He had evidence of a conspiracy among a number of Baltimoreans to assassinate the president-elect when he passed through the city. This corroborated alarming, but less definitive, information that Stone had already received from his own operatives. Stone thanked the detective and rushed to Scott's headquarters, where he made Bookstaver's revelation the key feature of his morning report to the general. Scott seemed concerned but unsurprised. He had been hearing similar warnings, but was now convinced that Lincoln could not get through Baltimore alive if he adhered to his high profile schedule, which called for a daytime reception there. When Stone insisted that the schedule be modified to a low-key night passage through Baltimore, Scott replied that Lincoln's "personal dignity" would not allow such a change. Personal dignity! An exasperated Stone explained to his commander that the assassination of the president-elect would leave the loyal states leaderless and likely precipitate civil war. These sentiments were in line with those of Scott, who then wrote a short note and directed him to deliver it to Senator William Seward, a strong Lincoln supporter and his future secretary of state. He finally tracked Seward down at the Capitol a little after noon. He gave

Scott's message to the senator and added one of his own. Seward left the Capitol immediately, found his son, Frederick, and gave him Scott's and Stone's notes along with one he hurriedly drafted. His instructions were for Frederick to take the first train to Philadelphia and hand the papers to Lincoln personally. He reached Lincoln in his room at Philadelphia's Continental Hotel at eleven P.M. and presented him with the documents. The elder Seward's note introduced Stone's memorandum and urged Lincoln to modify his schedule. The short note from Scott verified Stone's legitimacy, but the third, and most significant, message was Stone's transmittal of the information he had received from Bookstaver—specifically:

> A New York detective officer who has been on duty in Baltimore for the past three weeks reports this morning to Col. Stone that there is serious danger of violence to and the assassination of Mr. Lincoln in his passage through the city should the time of that passage be known—He states that there are banded rowdies holding secret meetings, and that he has heard threats of mobbing and violence, and has himself heard men declare that if Mr. Lincoln was to be assassinated they would like to be the men—He states further that it is only within the past few days that he has considered there was any danger, but now he deems it imminent—He deems the danger one which the authorities & people in Balt. cannot guard against—All risk might be easily avoided by a change in the travelling arrangements which would bring the Mr. Lincoln and a portion of his party through Baltimore by a night train without previous notice.[15]

Allan Pinkerton had already given Lincoln almost identical information gathered by his detectives. The president-elect asked Seward if he knew how Stone and Scott had learned of the plot, and if he had heard of Pinkerton. The young man answered no to both questions. Lincoln then explained that one report would not have worried him, but the results of two independent investigations gave him cause for concern. Nevertheless, he did not want to make a decision at that time. He would reveal his travel plans in the morning. After a night's sleep, he stated that he would fulfill his public obligations in Pennsylvania, but would follow a revised schedule developed by Pinkerton that called for transiting Baltimore in the small hours of the morning. The plan worked, and

Lincoln arrived unharmed in Washington on the morning of February 23. He lodged at Willard's, the district's most fashionable hotel and a headquarters for Unionists, where he polished his inaugural address and attended to other matters in preparation for March 4.[16]

Stone and Scott also were busy preparing for the inauguration, but their plans involved more than just protocol and oratory. While the president-elect had been delivered safely to the capital, the public swearing-in presented new challenges. There were still plenty of individuals in the city who would risk everything to prevent Lincoln from becoming president. One warning sent to him read:

> Caesar had his Brutus, Charles the First his Cromwell. And the President may profit from their example. From one of a sworn band of 10, who have resolved to shoot you in the inaugural procession on the 4th of March, 1861.[17]

In addition, the ceremony would attract visitors not only from the loyal Northern states but also from the less-trustworthy border regions. The inauguration would require the incoming and outgoing presidents to travel in a parade by open carriage from Willard's Hotel on Fourteenth Street up Pennsylvania Avenue to the Capitol, with the event itself taking place on a temporary platform erected on the building's east front. All of these venues presented opportunities for mischief.

As inauguration day neared, Scott and Stone marshaled their resources. On March 3, Scott held a final planning meeting at his headquarters. The company of regular army sappers and miners, which Stone had used as a threat in bluffing Captain Schaeffer, had finally arrived, as had some cavalry. Together with the artillery battery and the marines, these troops, which Scott believed to be more dependable than volunteers, now numbered about a thousand, more than during the dangerous days of January, but not enough to insure a trouble free inauguration. Scott and Lincoln would have to rely on Stone's militia. For some time, Stone had been considering how best to deploy his men for the event. Of course some would have to march in the parade, along with the sappers, miners, bands, and floats. Others would have more critical assignments. While visiting the Capitol, Stone had noticed that its two wings provided excellent fields of fire for covering the ceremony location. On March 4, therefore, he would station two riflemen at each window in the wings. Additional riflemen were to be positioned on rooftops lining the parade route with orders to shoot at anyone on foot or in a window who appeared threatening the presidential party. At Stone's suggestion, Scott also assigned regular cavalry to patrol the intersections between Willard's and the Capitol and artillery to guard the Treasury building and Capitol grounds. On the night before the inauguration, Stone received word that an attempt would be made to blow up the platform on which the ceremony was to be held. He immediately sent men to investigate and protect the site. Early the next morning, he detailed companies of the National Guard to line the Capitol steps and take positions under the stand. All was in readiness.[18]

The morning of March 4 was sunny, but with a cold wind. Lincoln and Buchanan emerged from Willard's Hotel after noon and boarded their carriage, the president-elect providing a better target because of his height. They were accompanied by Senators Alfred Pearce of Maryland and Edward D. Baker of Oregon, the latter a close friend of the president-elect. The procession began. What the men talked about, if anything, is not known. The incumbent was only too happy to be leaving the cares of the presidency, while his successor faced an unknown future, full of crises. They may have been distracted by the erratic movements of their mounted escorts. Stone, riding with this handpicked unit of volunteer cavalrymen and marshals, made "an apparently clumsy" use of his spurs to cause the cavalry mounts to bob and trot at an irregular pace. While these prancing horses made it difficult for spectators to get a good view of Lincoln and Buchanan as they passed by, he hoped that they would prevent any would-be assassins from getting off a fatal shot. After an uneventful passage up Pennsylvania Avenue the procession arrived at the Capitol, where marines and regular cavalry protected the inaugural party as it entered the building. Stone made his way to a Capitol window from which he could view the proceedings as they unfolded on the steps below. Several plain-clothes policemen mingled in the large crowd in attendance with orders to "strike down any hand that might raise a weapon." At last, the cannon boomed signaling that Abraham Lincoln was now president of the United States. While Scott had the responsibility for the inauguration he did not participate in the procession or ceremony, but viewed the event from his carriage on a low hill near one of the artillery emplacements. Observers at the time gave Stone most of the credit for managing the day's security.[19]

With Lincoln's inauguration, Washington entered into a period of surreal calm. Stone called it "the lull which preceded the cyclone." Lincoln had stated that the national government would not provoke an armed conflict with the seceded states, but at the same time he had promised to maintain the Union and protect federal facilities in the South. Washingtonians knew that the status quo could not last, but few believed that the country would devolve into civil war. Past crises had been avoided by compromise, and certainly the new president would be able to pull a dove of peace out of his stovepipe hat. The capital's military leadership, however, was not as sanguine about the prospects for a peaceful settlement. Scott and his staff sought to place the district on a wartime footing. In the meantime, all eyes were on the attempts to resupply Fort Sumter in Charleston Harbor and on Virginia, which had not yet decided on the question of secession. If she left the Union, Maryland would likely follow, and Washington would find itself surrounded by potentially hostile powers.

Scott and Stone recognized the capital's precarious position and did their best to strengthen its defenses without provoking Virginia and Maryland into secession and war. Stone continued organizing and training militia units. His drilling was incessant, for the colonel anticipated that "the explosion [of civil war] might occur at any moment." By mid-March, he had some 3,500 militiamen under arms, and urged the War Department to muster them into United States service. This idea, however, was initially rejected as unnecessary, and as late as April 5, Scott believed that calling up the militia to defend the capital was not necessary. Stone was not so sure. He saw the Southern threat as real and imminent. That day he met with Secretary of State Seward to discuss the security of the capital and later wrote him a letter relaying Confederate threats to seize Washington that his sources had reported to him. (Mrs. Jefferson Davis was said to have remarked that she would be receiving her old friends at the White House in June.) As the Sumter crisis deepened over the next couple of days and Virginia edged closer to joining the Confederacy, even the War Department realized the need for an effective fighting force in the district. On April 8, Scott requested that Stone determine the number of "reliable" volunteer soldiers. The next day the order came to muster in an initial four militia companies as United States Volunteers in Executive Square, near the White House. The process did not go smoothly. At first, some men, despite Stone's weeding efforts, refused to take the oath of allegiance if it meant they might possibly fight against Virginia, Maryland, or other Southern states. Others refused the oath since it did not indicate how long they would be called on to serve. And still others declined because they feared they would be sent out of the Washington city limits. Defending their own homes and businesses was one thing, participating in a long campaign far away was quite another matter. Finally, the assistant adjutant general, Major Irvin McDowell, promised the assembled troops that they would be stationed only within the district unless they volunteered to go elsewhere, and that the term of their federal enlistment would be no more than three and a half months. Thus assured, most of the four companies were mustered in. The pro-Southern militiamen who did not take the oath had to surrender their weapons and uniforms immediately, and were told to disperse.[20]

The federalizing of the district militia continued for the next few days, and by April 12, the day of the attack on Fort Sumter, fifteen companies had been mustered in. The War Department then called up an additional eight companies. Scott was confident that this would provide him with enough soldiers to handle any threats to the capital until regular army and volunteer units from other states arrived in about a week. He miscalculated. While Lincoln's April 15 call for seventy-five thousand volunteers to put down what he called an "insurrection" galvanized Northern patriotism, it had the effect of tipping Virginia to the Confederacy. Secessionists in the Old Dominion argued successfully that Lincoln intended to march his volunteers through the state in an invasion of the South. On April 17, Virginia's leaders responded with an ordinance of secession, which they submitted to the voters. Although the results of the plebiscite would not be known for several weeks, its outcome was certain. After April 17, Washington's defenders had to be ready to repel an invasion from across the Potomac River. Slaveholding Maryland became an even greater problem. The state did not secede, but pro-Southern activists stirred up a wave of popular sentiment against the Union, particularly in Baltimore and the surrounding region. Secessionists tore up railroad tracks, damaged bridges, and otherwise disrupted transportation and communication lines serving Washington. At the same time, even if he had had the troops available, Scott realized that an aggressive move against the secessionists might well push Maryland into the Confederacy. With both states now considered enemy territory, the capital was surrounded and faced the very real possibility of a siege.

Adding to Scott's dilemma was the delay in getting reinforcements. Some federal units in the North made preparations to move on the capital, but they seemed to lack a sense of urgency. Volunteer regiments from the Northern states, on the other hand, were anxious to defend the district, but the situation in Maryland prevented most from getting there in a timely manner. By the time of Virginia's secession, some companies of Pennsylvania volunteers had made it through, although they arrived without weapons, save for a few sabers. On April 19, the 6th Massachusetts Volunteer Regiment was attacked by a pro-Southern mob as it passed through Baltimore. In the ensuing melee, shots were fired leaving four soldiers and some attackers killed and dozens more on both sides injured. Bands of secessionists also damaged the tracks and placed impediments in the way of the train carrying the Massachusetts men. While their train finally arrived in Washington late in the day, their experience made it clear that the most direct route from the North to the capital was problematic at best. The Pennsylvania and Massachusetts troops joined the few regular units that had been in the city since the inauguration and the marines. While the presence of these additional blue uniforms may have buoyed the spirits of patriotic Washingtonians, they alone were not enough to mount an adequate defense of the city. Scott would still have to rely on the district militia.

On April 16, Stone was appointed to command the militiamen who had been sworn into federal service. Over the next couple of days, he organized these 3,500 citizen soldiers into battalions of four companies each and made plans to deploy them to suppress an insurrection or repel an attack. At the White House he posted double sentries behind bushes and other places of concealment so that, "a person entering by the main gate and walking up to the front door . . . could see no sign of a guard," so that it did not seem as though the president needed heavy security. The visitor, however, "was under the view of at least two riflemen standing silent in the shrubbery and any suspicious movement on his part would have caused his immediate arrest."[21]

In addition to the White House, Stone assigned overnight guards at the Treasury, State, War and Navy Department buildings in Executive Square, Scott's headquarters in the Winder Building, the Capitol, city hall, and the General Post Office. He also wasted no time in guarding the approaches to the district. Of greatest concern were the crossings over the Potomac—the Navy Yard Bridge, the Long Bridge on the Alexandria Road, and the Chain Bridge just upstream. These spans connected Washington with Virginia and were the most likely invasion routes. With Scott's approval, Stone placed detachments of soldiers at all four to prevent unauthorized movement from across the river and contact with the enemy. The Long Bridge was deemed the most critical, since it led to the heart of the city. There, in addition to infantry pickets, he detailed dragoons and a cannon. Even with this extra force, the manpower at the Long Bridge did not number above fifty at any one time. Other pickets were stationed at the Seventh Street Road and Fourteenth Street entrances to the city from Maryland, and at Tennally Town in the northwest portion of the district, where the roads from Poolesville and Rockville, Maryland, converged.[22]

Anticipating a siege, Stone moved quickly to secure the district's food supplies. He surveyed stores of grain in the city and found only a three-day supply on hand. Flour mills in nearby Georgetown, however, were said to hold 10,000 barrels, so he posted guards to keep an eye on the plants. When he learned a couple of days later that the flour was to be shipped away (probably to the Confederacy), Stone contacted the new secretary of war, Simon Cameron, and together they arranged to seize and transport all 10,000 barrels to Washington. The supply was stored in the Capitol, Treasury Building, and post office. It would prove critical when the city was, indeed, cut off.[23]

Stone also decided to capture the steamboats plying the Potomac and Aquia Creek lest they fall into the hands of the Virginians. One of the vessels, the *St. Nicholas*, he suspected was preparing to carry supplies to the rebels at Alexandria across the river. In order to nab the conspirators, he asked Secretary Cameron if he could take over the Washington office of the American Telegraph Company to prevent word of the impending seizure from being broadcast. His request amounted to a form of military censorship on outgoing messages. When the secretary hesitated, Stone insisted, "Nothing should, from this time on, go over the wires from here unless approved by the War Department. We are in a state of war and should act accordingly."[24] Cameron finally consented, as did Secretary of State Seward, providing that Stone take full responsibility for any consequences. Stone readily agreed, rounded up the *St. Nicholas* suspects, and with a squad of volunteers seized the telegraph office, embargoing all outgoing communication. Late the following day, when word of the Baltimore attack on the Massachusetts regiment reached Washington, journalists rushed to file their stories, only to find the telegraph office closed and guarded by soldiers. The frustrated reporters hunted up Stone, who sloughed off his role in the closure by saying that he had been ordered to do so. Orders were orders and he could not change them. Getting nowhere with Stone, they went to Seward, who neither confirmed nor denied that he had advised the government to censor the press. The situation festered for a day until an agreement was worked out to allow regular business and officially authorized messages to go out. Military information, however, would continue to be detained and reviewed by a censor appointed by the War Department.[25]

Between arranging the capital's defenses and thwarting rebel plots, Stone maintained a hectic schedule. He reported to General Scott in the mornings and afternoons, and spent most of the day checking on the guards and pickets stationed about the district. This left him precious little time for rest, as he later recalled:

> The only sleep which I could snatch was taken in a carriage while driving from one picket to another. The drive from Long Bridge to Chain Bridge would afford me a nap; that from Tennally Town to the Fourteenth Street picket, another; that from Seventh Street to Benning's Bridge, yet another; and that from Benning's Bridge to the Capitol, one more.[26]

Meals likewise were taken on the fly. From the Capitol he would head to the White House to inspect its guards, and then on to the Treasury Building by sunrise to catch a bit of breakfast and begin preparing his morning report for his commander.

On April 21, Stone's routine was interrupted by an unusual request from Scott. It was the day after federal forces had abandoned and torched the Gosport Navy Yard opposite Norfolk, Virginia. By then it also had become clear that the rail line through Baltimore would be unusable for some time, and word had reached Washington that the arsenal at Harper's Ferry had fallen, although most of its stand of 15,000 rifles, equipment, and machinery had been destroyed prior to falling into Confederate hands. Amid all of this and other bad news, Stone made his regular morning call on the general. He had not yet begun to give his report when Scott looked up from his desk and abruptly asked—ordered—Stone to alter his schedule and meet him for dinner that evening at his quarters. Dismissed.

Stone reported to Scott's residence promptly at half past four and was shown into the dining room where the general was seated. Burdened with responsibilities that would have taxed the constitution of a much younger man, the old warrior appeared careworn and distracted. The two men downed their soup without a word being said. The silence continued into the main course and was broken only when Scott asked Stone to carve the chicken, which he did, giving the general his favorite portion. Once again, silence descended on the dinner table. Finally, a telegram arrived which caused Scott to remark dispiritedly, "Colonel Stone, we have fallen on evil days." The message was from an old friend asking if Scott would switch sides and join the South. As they poured their after dinner sherry, Scott raised his glass as if to propose a toast. Instead, Stone recalled that he looked him straight in the eye and said:

> "Gosport Navy Yard has been burned!" I replied quietly: "Yes, General!" He continued: "Harper's Ferry Bridge has been burned!" Again I replied: "Yes, General." Again he spoke: "The bridge at Point of Rocks was burned some days since!" I replied: "Yes, General." He continued: "The bridges over Gunpowder Creek beyond Baltimore have been burned!" I still replied: "Yes, General." He added, "They are closing their coils around us, sir!" Still I replied in the same tone: "Yes, General." "Well, sir," said the General, "I invited you to come and dine

with me to-day because I hoped you could listen calmly to that style of conversation. Your very good health, sir." He drained his glass while I bowed and followed his example.[27]

Scott's ramble broke the tension and the two got down to business. He asked Stone how long he thought the district could hold out. Stone responded, ten days. By that time reinforcements surely would have arrived. When asked about his defensive perimeter and the number of troops he could rely on, Stone stated eighteen miles and 4,900 men, including regular army, volunteers, and marines. Scott, concerned about the thin line of defense, then blurted, "You will have to fight your pickets!" This ran counter to prevailing military practice of using pickets to gather information on an advancing enemy and not as a primary line of defense. True, Stone countered, but his plan relied on the likelihood that the rebels would not have sufficient strength to move on more than one entrance to the city at a time. The commotion from that location would alert other pickets who could converge on the point of attack. If they could not hold, they would fall back, firing, to three strongholds: the Capitol, where flour was stored and 250 volunteers were stationed; city hall, which had a view of the city and ready access to the General Post Office, where flour also had been cached; and Executive Square, where the Treasury Building with its thick walls, water supply, and flour storage would be the strongest point. The few artillery pieces he commanded would move about the city to be employed wherever they were needed the most. From these locations, Stone believed he could fend off the attackers until help arrived. "This is all we can do, and what we can do must be done."[28]

Scott pondered the plan and the map of the district Stone had spread out on the dining table to illustrate it. Finally he said the plan was generally satisfactory, "It is all that can be done." But, he added that holding all three strongholds was impractical. There should be a progressive withdrawal, first from the Capitol, where the only things to burn were the libraries of the Congress and Supreme Court. In abandoning the Capitol, however, he did not believe that, "American soldiers, even in rebellion, are yet capable of destroying public libraries and the archives of courts of justice."[29] The city hall would be the next to be vacated. When Stone objected that the site offered a "commanding position," Scott replied, "It is a pity to abandon such a commanding position, as you say, my young friend. But we must act according to the number of troops we have with which to act. All else must be abandoned, if necessary" to defend Executive Square.[30] The ultimate stronghold would be the Treasury Building, and, perhaps the State Department. Scott then directed Stone to gather the seals of all of the agencies of government and deposit them in the Treasury so that they would not fall into the hands of the rebels who could use them to forge documents and otherwise create confusion among offices distant from Washington. Scott concluded that if worse came to worst, and the Treasury did become the government's "citadel" for a last stand, the president and cabinet must be brought there. "They shall not be permitted to desert the capital."[31] There would be no government in exile for the United States.

Thus was developed the worst-case scenario for holding Washington. A few days later, the gist of Scott's and Stone's after-dinner conversation appeared as General Order Number Four. Regular army officers were assigned to command the key locations in the event of an emergency. Having done so much to develop the defense strategy, Stone, although technically a volunteer soldier, claimed the most important stronghold for himself—the Executive Square.[32]

The next few days would be the most critical for Washington. Reinforcements were said to be on the way, but no one knew when they would arrive. Although the battered 6th Massachusetts Regiment had made it through Baltimore, it was unlikely that any others would follow. An incorrect rumor circulated that Virginia had closed the mouth of the Potomac. That left ferrying troops to Annapolis and then using a branch rail line as the most likely path for reinforcing the capital. Indeed, Northern soldiers had landed at Annapolis, but the tracks out of that town had been torn up, and marching to Washington would require passing through hostile territory under constant threat of ambush. In the meantime, while Stone's volunteers patrolled the streets and manned the perimeter, secessionist partisans openly played "Dixie" and shouted, "Hurrah for Jeff Davis!" Their taverns and meeting halls were crowded and raucous. Unionists, on the other hand, went about their business with a cloud of gloom hanging over their heads.

Many left town or were preparing to evacuate. Something had to be done, and quickly.

Scott did not have the manpower to force an opening of the rail line through Baltimore and lacked the ships to break a blockade of the Potomac.[33] The Annapolis branch would have to be opened. Early on the morning of April 23, therefore, Secretary of War Cameron sent orders through Scott for Stone to seize and hold the Baltimore and Ohio Railroad station on New Jersey Avenue between D and C streets, about two blocks north of the Capitol. Stone immediately directed a battalion of volunteers to meet him there. They entered the building and found the place deserted, save for the stationmaster. Only one derelict locomotive stood in the yard. As Stone set his men searching for rails or any other equipment that could prove useful, Scott and Secretary Cameron arrived and told him to take possession of any locomotives and rolling stock that might show up. Scott issued an additional order for Stone to move his men up the line to Annapolis Junction and determine if the rumors were true that the branch to Annapolis had been torn up. While he was figuring the best way to do this, a locomotive with a baggage car and two coaches steamed into the station. Some luggage and freight was thrown onto the platform from the baggage car and three men hurriedly detrained, collected their bags, and hustled away. When the train began to back out, Stone stepped forward and ordered the engineer to stop. The engineer shouted that he intended to return to Baltimore, but as he turned around he saw the tracks behind him covered with soldiers. When a freight train arrived a short time later, it too was detained. Stone then ordered a special train to be made up and asked Scott for additional instructions, but the general told him to use his own discretion. Given a free hand, Stone was ready to dispatch the train to Annapolis Junction. His troops, however, had been promised by Major McDowell that they were sworn into federal service only in the District of Columbia. They could volunteer to go elsewhere, but could not be compelled to do so. He commanded the nearest company to assemble in front of him and declared:

> Soldiers, you have been mustered into the service for duty in the District of Columbia only. I do not claim the right to send you out of the District without your consent, and will not do so, but now I want 200 men to board that train yonder, to go wher-

ever I say, to do whatever they may be ordered to do under the command of any officer I may designate. This service is important for the Government, and it may be dangerous. All in the company who wish to go under these conditions, step one pace to the front![34]

All but the bugle boy advanced. When he delivered a similar address to a second company, the entire unit volunteered to go. Breathing a sigh of relief, Stone placed Captain William B. Franklin, a regular army officer who had graduated first in the West Point class of 1843 and was lately supervising engineer for the construction of the Capitol dome and architect of the new Treasury Building, in command of the mission.

Before the train could leave, however, it was discovered that during the organizing of the expedition, the locomotives had been sabotaged. The Baltimore and Ohio mechanics had disappeared, and none of Stone's volunteers possessed the skills needed to effect the repairs. The colonel then hit upon the idea of asking if any of the men of the 6th Massachusetts then billeted at the Capitol were mechanics. Within a few minutes six troopers arrived carrying their tools. In short order they had a locomotive in running order, but in the confusion, the engineer had left. Stone questioned the stationmaster and learned the whereabouts of the wayward trainman. He sent a squad of soldiers to retrieve him, but once back at the locomotive, the engineer and his fireman refused to take the train to Annapolis Junction. Stone made it clear that they had no choice. He would post two riflemen on the engine platform with orders to shoot them if they attempted to disable the locomotive or desert. When the engineer continued to protest that if any of the bridges or culverts had been weakened he could die an innocent man, an exasperated Stone replied that such an event would be unfortunate, but he did not believe that, "any bridge or culvert would be made dangerous by the rebels without you being informed of it." After all, he had been anxious to take his train on those same tracks to Baltimore just a short time earlier. The colonel continued, "and if knowingly you wreck this train, you will . . . be justly shot on the spot." With that, the reluctant crewmen climbed onto the locomotive and the train departed. Stone later recalled that he "felt sure of the best efforts of the engineer and the fireman."[35]

Captain Franklin and his two companies proceeded up the main line of the Baltimore and Ohio to Annapolis Junction, leaving guards at key locations along the route. When they arrived, they found the tracks on the branch to be either removed or torn up for as far as the eye could see. A trained engineer, Franklin set about developing estimates of the amount of rail on sidings and spurs that could be salvaged to replace what the secessionists had damaged or hauled away. When his scouts returned with the news that volunteer troops under the command of Brigadier General Benjamin Butler were, indeed, in Annapolis and had begun making temporary repairs to the damaged tracks and moving toward Annapolis Junction, Franklin returned to Washington to report to Stone.[36]

Before leaving to communicate the results of Franklin's reconnaissance to his superiors, Stone dispatched another train to Annapolis Junction to keep the line open and monitor Butler's progress. Over the next couple of days, similar trains would shuttle between Washington and the junction. When he met with Scott, Stone found that Thomas Scott, first vice-president of the Pennsylvania Railroad, had been appointed to manage the rail lines and telegraph facilities that had been commandeered. General Scott was pleased with the report, and when Stone presented him with Franklin's estimates on reconstructing the tracks to Annapolis, information that Tom Scott had been requesting from the War Department, the delighted general chortled: "Anticipated Again! Oh! These rascally regulars!"—adding quickly that he used the term "rascal" because he loved them so.[37] Although the siege of Washington had not yet been lifted, there was hope on the horizon.

By April 25, the work of rebuilding the Annapolis line had been completed, and the first troops of Butler's command arrived that afternoon. When the train, covered with men of the 7th New York Infantry, steamed into the Baltimore and Ohio station, the militia units present let out a joyful shout, followed by the 6th Massachusetts, and most of the loyal population of the district. The 8th Massachusetts Infantry arrived twenty-four hours later, and would be followed by the 5th Pennsylvania, 5th Massachusetts, 1st Rhode Island, 12th New York, and 25th New York regiments. By the end of the month, the ranks of volunteers ready to defend the capital had swelled to almost ten thousand. Stone, who had been the district's rallying point in its darkest hours, later summed up the result of his efforts when he wrote, "from that day [April 25] on, each train sent out came back to us laden with volunteers from the Northern cities, and Washington was soon too strong for attack by any force which its enemies could then send against it."[38]

GENERAL STONE

THROUGHOUT the month of May 1861, Washington took on more the appearance of an armed camp than a political capital as regiment after regiment of volunteers and regulars arrived in the city. Military reviews became an almost daily occurrence, and loyal Washingtonians reveled in the knowledge that, for the foreseeable future, there would be no rebel insurrection or invasion. Amid the sea of blue coats and colorful *zouave* uniforms, however, the officer who had done so much to defend the district over the past four months watched the parades from the sidelines. Colonel Stone still served as inspector general for the District of Columbia and commanded its militia units, but the importance of these assignments diminished with each arriving troop train. It was clear that a new army was being formed, and the District of Columbia Militia would be just one small and relatively inconspicuous component. As the right-hand man of the general-in-chief, Stone might have expected to assume a high-echelon position in the construction and operation of the Union's war machine. He must have been surprised and disappointed, therefore, when he did not receive a brigadier's star right away. Instead, on May 14, he was mustered into the regular army as a colonel and given command of the 14th Infantry Regiment, a unit that would not be organized for months. In the meantime, he had to be content to drill his militiamen, supervise his pickets, and await an opportunity for advancement.[1]

On May 22, the result of the Virginia secession plebiscite became official. The Old Dominion would be joining the Confederacy. This announcement prompted Scott, who had been postponing a move against Virginia so as not to add fire to the secessionist cause, to launch an attack across the Potomac. The target would be Alexandria, a sleepy commercial town and center for the slave trade opposite Washington. Three columns of infantry and cavalry would cross the Long Bridge and secure the countryside around the town while a regiment of "Fire Zouaves" from New York, led by the dashing and popular Elmer Ellsworth, took a steam sloop and transports to invade Alexandria proper. Stone was given one of the columns consisting of militiamen, whom he again had to convince to serve outside the district. Just after midnight on the twenty-third, he took a detachment of mounted militia across the Chain Bridge upstream from the main point of attack. Once on the Virginia side, Stone and his men galloped to the Long Bridge, where they surprised and dispersed the rebel guards, who had orders to burn the structure in the event of threatening Yankee moves, thus clearing the way for the main body of troops to move across unmolested. By this action, he became the first Union officer to lead an invading party into enemy territory. As the other regiments streamed into Virginia, Stone sent his men ahead to clear the roads and report any organized resistance. None materialized, and by dawn Union troops controlled Alexandria. Only a few shots were exchanged before the defenders either fled or had been captured. Colonel Ellsworth was the only Northern casualty, having been killed by a secessionist innkeeper after removing a Confederate flag flying over the hotel. Grieving Unionists quickly elevated Ellsworth to the status of first martyr for the Union. As the flurry of eulogies began, Stone's ride, which had enabled the almost bloodless invasion, was quietly forgotten by most Washingtonians.[2]

The army, however, did not ignore Stone's initiative in the march on Alexandria. Initially, he was given charge of the captured town, but when Colonel Joseph K. F. Mansfield, commanding the

Department of Washington, learned that only Stone knew the deployment of pickets and other defenses around the district, he quickly recalled him. On May 28, Stone received new orders transferring him back across the Potomac to the Department of Northeastern Virginia under General Irvin McDowell. He was given a brigade, but three days later, McDowell removed him from this duty and assigned him to a mission devised by his old commander. Winfield Scott still maintained a high opinion of Stone for his work in defending the capital and president, and directed him to take a "strong brigade" up the Potomac River to secure the important crossing at Edward's Ferry and stop the smuggling of supplies from Maryland into Virginia. While he mentioned the possibility of capturing Leesburg, Virginia, he left that choice to Stone's "well known discretion," cautioning the colonel against undertaking anything that could endanger his brigade. Scott also envisioned that Stone would eventually join with the army commanded by Major General Robert Patterson, a veteran of the War of 1812 and the Mexican War, in operations in western Virginia.[3]

The "strong brigade" numbered some 2,500 men consisting of the 9th New York State Militia, 17th Pennsylvania Infantry, and 1st New Hampshire Regiment, along with four battalions of the District of Columbia Militia (which, of course, had to be cajoled to leave the city), a company of the President's Mounted Guard, Company H of the 2nd Cavalry, and a light company of the 5th Artillery. The cavalry and artillery units, along with the New York, Pennsylvania, and New Hampshire regiments, left Washington on a calm, hot June 10, marching northwest through Tennally Town over a bad road, "full of stones and very dusty," to Rockville, Maryland, which they reached by sundown. Stone and most of the District of Columbia Militia followed, spending their first night in Tennally Town. The following day, a battalion of District of Columbia Militia was sent up the River Road, which roughly paralleled the Chesapeake and Ohio Canal, while the remainder of the column proceeded on to Rockville. There, someone, perhaps Stone, gave a name to the campaign—the Rockville Expedition. The mood was upbeat, even festive, as the historian of the New York troops recalled: "The men tramped gaily along, exerting themselves to the utmost, happy in the belief that the end would bring them nearer the enemy, whose prowess they had not then learned either to fear or respect."[4] While their comrades enjoyed the outing, men of the 5th District of Columbia Militia Battalion stayed behind in Washington. They would not leave until June 12, moving up the Chesapeake and Ohio Canal in two boats full of provisions and supplies.

The arrival of Stone's brigade at Rockville had the desired effect of encouraging loyal Marylanders in the region who had been intimidated by secessionist gangs. As a result, pro-Union candidates triumphed there in a June 13 special election, as they did throughout the state. While in Rockville, Stone learned of Confederate activities along the Potomac, including crossing the Potomac near Edward's Ferry and draining a portion of the Chesapeake and Ohio Canal. He had to move, and quickly. Before leaving, Stone attempted to disguise the expedition's true objectives by making conspicuous inquiries as to the best route to Frederick, Maryland, sensing that this information would be communicated quickly to the Confederates by die-hard Southern sympathizers. Instead of marching to Frederick, however, Stone and the bulk of his command headed toward the Potomac River, marching to Darnestown, where they designated their bivouac, "Camp Stone," and on to Poolesville, roughly opposite Leesburg. Stone, meanwhile, went on a scout to Edward's and nearby Conrad's ferries but found no signs of rebel activity. He returned to Poolesville, which became his headquarters and would remain so off and on for the next several months. From Poolesville, Stone sent artillery and infantry to occupy the principal fords and crossings up and down the Potomac.[5]

For the next few weeks, his men skirmished with rebel pickets who made attempts to dislodge them from their positions, but little harm was done to either side. Stone reported one such encounter at Conrad's Ferry about five miles upstream from Leesburg:

> The enemy opened fire on the guard . . . this morning about ten o'clock. The point was and is occupied on our side by five companies of the New Hampshire First Infantry. The enemy were reported to have three cannon, but in a careful examination, I was unable to discover more than one 6-pounder field piece. They amused themselves by firing some twenty shots, apparently at the staff on which the New Hampshire troops had

raised the national colors. No damage whatsoever was done to our men by the firing.[6]

Once they had ceased firing their six-pounder, the rebels went down to the riverbank and hurled insults across until their voices gave out. While this engagement may have seemed inconsequential and even amusing to his troops, Stone sensed a more devious motivation behind the rebels' actions. He suspected that the shelling was a ruse to draw his attention from a possible Confederate crossing in force of the Potomac upstream. Consequently, he quickly redeployed his men to meet the possible attack. It never came.[7]

A seemingly more serious action occurred the next day when an estimated eight hundred to one thousand Southern troops gathered at Edward's Ferry near Leesburg and tried to cross Goose Creek where it joins with the Potomac preparatory to an attack into Maryland. To accomplish the crossing, the rebels sought to use a derelict ferryboat. Stone, however, had given his artillery battery orders to sink the vessel if the enemy tried moving it. Accordingly, when the rebels pushed the boat away from the bank, the Union artillerists fired a spherical case shot from their twelve-pound howitzer. The shell exploded over the ferry, covering it with "a shower of bullets and fragments." In his report, Stone noted, "The effect was excellent. The horse of a mounted officer leaped overboard and the boat was rapidly drawn back to the shore."[8] The Confederates then lined up and commenced firing across the Potomac, and, once again, the Union men employed case shot to disperse their foes. So went the war along the Potomac during the innocent first months of the conflict.

When not fending off halfhearted aggression by the rebels and occasional bushwhackings by pro-Southern Marylanders, the Rockville Expedition efficiently accomplished its mission of securing Edward's Ferry and other crossings and stemming the smuggling of contraband into Virginia. It also monitored enemy troop movements and was able to open the lower portion of the Chesapeake and Ohio Canal, allowing grain and coal supplies to reach Washington. His opponents across the river, however, could not figure out what the Rockville Expedition was really up to. When word of Stone's advance out of Washington reached the

Confederates they assumed it meant an invasion of Virginia. Colonel Eppa Hunton of the 8th Virginia Infantry was sent to prepare the defenses of Leesburg and surrounding Loudon County. When the first Union troops appeared on the Maryland shore, he increased the guards at the fords and crossings. (To Stone, this seemed as though the Virginians were preparing their own attack on Maryland.) On June 8, even before the Rockville Expedition had left the capital, a subordinate reported to Hunton that a large Union force was going to cross the river at Edward's Ferry. He ordered his men to prepare for the onslaught, but nothing happened. When he reached the supposed invasion site he learned that "there had not been a single effort to cross; no preparation to cross; and evidently no idea on the part of General Stone of crossing the river."[9]

Hunton's assumption was only partially correct. Throughout the Rockville Expedition and beyond, Stone devoted serious thought to invading Virginia and taking Leesburg. Scott had allowed him to be "governed in other operations by information you gain," and the information he was gaining indicated that the rebels were frightened by the Rockville Expedition's appearance along the Potomac. Stone's first inclination was to capitalize on the enemy's supposed timidity, cross the river, defeat whatever rebels might be around, and seize Leesburg. But a lack of artillery caused him to stay on the Maryland side. Following Hunton's abortive attempt at crossing Goose Creek, Stone reported to Scott that he could, if he wished, occupy Leesburg with few casualties. For some reason, however, he concluded that even such a slight risk would not justify an attack. Instead, Stone extended his defensive line to crossings upriver from Conrad's Ferry. From these positions they could observe rebel movements in Virginia and intercept more smugglers.[10]

While Leesburg remained a tantalizing prize, Stone attempted to find out the location of General Patterson's Pennsylvania column, with whom he was to coordinate. If he had Patterson's support, an assault across the Potomac would be quite feasible. Before making such a move, however, he also had to determine if Confederate general Joseph E. Johnston and his army were, indeed, occupying Harper's Ferry, as had been rumored. If that were the case, Stone concluded that his line was overextended and vulnerable to an attack. He hoped for assistance from Patterson, but although the general was supposed to be moving on Harper's Ferry, the Pennsylvania column was nowhere to be found.

By June 23 Stone had just about changed his mind about Leesburg and decided that if he could determine that Patterson had compelled Johnston to retire and held Harper's Ferry, he would occupy Loudon County. Such a move would offer protection to the Union army's right flank in Virginia and free the Chesapeake and Ohio Canal above Edward's Ferry from rebel harassment, allowing it to become fully operational. With no word from Patterson, however, Stone had to consider the possibility that Johnston's force was still present at Harper's Ferry and dropped the plan. Patterson, meanwhile, had developed his own plan to join with Stone in the capture of Leesburg, but this scheme, too, was abandoned on June 30 when Scott, fearful of Johnston's presence in western Virginia, ordered Patterson to remain opposite Harper's Ferry. Stone would march to meet him there. Of course, Stone knew nothing of this until July 1.[11]

Despite repeated attempts to contact Patterson, Stone remained in the dark. On June 20, a scout had brought him word that no troops, Union or Confederate, were at Harper's Ferry. At this point, he decided to take matters into his own hands and find out for himself just what was going on. Accompanying him on an upriver reconnaissance would be his adjutant, Captain William S. Abert, his aide-de-camp, Captain Stewart, and a dozen mounted troopers. The party left Poolesville at midnight of the twenty-fourth and by three A.M. had moved beyond the forward Union pickets. At dawn they arrived at Point of Rocks, where they found a few rebel sentries on the Virginia side but no sign of Patterson or other Union soldiers. Stone then sent Abert forward to Sandy Hook, across the Potomac from Harper's Ferry. There he encountered an engineer officer from the Pennsylvania column who, unfortunately, could offer little current information on Patterson's location or movements. Local Unionists received Abert warmly and informed him that Johnston had abandoned Harper's Ferry, but was rumored to be leading a large force toward Leesburg with the intention of attacking General McDowell's right flank. Stone returned to his headquarters and quickly passed this information on to Scott. He also stated that he would occupy Point of Rocks and could do the same at Sandy Hook if more artillery and an additional infantry regiment were assigned to him.[12]

Stone waited for a response from Washington for the next few days, but none came. With Patterson's movements still unclear, he had no choice but to stay put in Poolesville. Finally, on July 1, he received Special Order Number 109 instructing him to send his mounted troops, artillery, and "such of the District of Columbia Volunteers as he may desire to return" back to the capital. He then was to proceed with the remainder of his infantry to join with Patterson's army wherever it happened to be. Stone realized this order meant the end of the Rockville Expedition as an independent operation, and reluctantly complied. He dispatched his cavalry, artillery, and a portion of the District of Columbia troops back to Washington, and with his reduced brigade commenced the trek to Patterson's "supposed position."[13]

By July 4, Stone had reached Point of Rocks, and that afternoon, the 9th New York Militia marched to his temporary headquarters and requested his presence for a "grand festival." He appeared and delivered a "short but feeling" speech, after which the regiment sang a number of patriotic songs. The music was interrupted, however, by word that there had been skirmishing at Harper's Ferry between a handful of rebel pickets and Stone's forward units. The 9th New York hurried off to support their comrades, abruptly ending the impromptu Independence Day ceremony.[14]

The exchange at Harper's Ferry lasted only an hour and resulted in one Union soldier killed and another wounded. The enemy retired to the outskirts of the town, but their presence concerned Stone, since he had not yet been able to contact Patterson. Were these men the advance guard of a much larger force that could threaten the Union position in western Virginia and Maryland? Or had they been left there to annoy the Yankees and report to Johnston on Stone's and Patterson's movements? Or were they just stragglers? While he was pondering these possibilities, and considering the occupation of Harper's Ferry or a raid to cut the rail line connecting western and eastern Virginia, orders finally arrived from Patterson's assistant adjutant general, his old West Point classmate and Mexican War comrade, Major Fitz-John Porter. Stone was to proceed to Harper's Ferry, and from there to the Charlestown, Virginia, area to join the main column.[15]

At last! Patterson had materialized from the mists of western Virginia, and Stone now had a destination. He wrote back to Porter that he would attempt to obey the general's instructions, but crossing the Potomac and moving through enemy country without cavalry or artillery would be slow and hazardous. This response did not sit well with Patterson, who had Porter send a sharp note explaining that it was urgent that Stone join his column at the earliest possible moment. He should reduce his baggage as much as possible, including abandonment of his tents, and hire or even requisition under promise to pay teams to move the train if necessary. By no means was he to dawdle. Stone must have seen an irony in the message. For over a fortnight he had been trying to contact Patterson, with no response from the general, and now he had to pack up almost at a moment's notice and scurry through possibly dangerous territory. While he may have grumbled over Patterson's orders, he took heart in a dispatch from Scott in which the old general praised him for the Rockville Expedition's success in opening the lower Chesapeake and Ohio Canal and stopping the flow of contraband from Maryland to Virginia. Scott also apologized for not sending Stone reinforcements and artillery, thereby denying his Rockville Expedition the chance to take Leesburg and inflict other damage on the rebels across the Potomac.[16]

Stone's brigade joined Patterson at his new headquarters at Martinsburg, Virginia, on July 8. There it was reorganized as the 7th Brigade of the Army of the Department of Pennsylvania. Initially it was attached to a division commanded by William Keim, a major general of the Pennsylvania Volunteers, but was soon transferred to the division led by New York's major general Charles Sanford. Sanford, who had overseen the capture of Alexandria, knew of Stone's ability and designated his brigade to play an important role in an attack on Johnston's force at Winchester, Virginia, the next day. The newly arrived troops, however, were fatigued from their march. Patterson deemed them unfit for combat and cancelled the attack. Instead he held a "council of war" with his senior officers, the result of which was an alternative plan (with which Stone agreed) not to advance on Winchester, where Johnston was believed to have an overwhelming advantage in men and artillery, but to attempt to outflank him by moving southeast to Charlestown. Once this decision had been made, Patterson still lingered at Martinsburg for another week, more than enough time for the new Seventh Brigade to rest up. This must have seemed odd to Stone, since the old general had been so adamant that he rush to join him.[17]

On July 13, an increasingly apprehensive Scott told Patterson that if he could not defeat Johnston, he should at least detain him and prevent him from uniting with the main body of the Confederate army. This order was to correspond to General Irvin McDowell's advance on Bull Run and the key railroad junction at Manassas, Virginia, due to begin on July 16. Patterson complied cautiously, believing that his foe had twice the number of men than he actually commanded. He advanced to Bunker Hill a few miles from Martinsburg on July 15, but the next day, finding the roads to Winchester obstructed, he fell back to Bunker Hill and then made a left turn toward Charlestown. As late as July 18, Patterson believed he had accomplished his assignment of keeping Johnston occupied. But such was not the case. Patterson's attempted flanking move proved to be just what Johnston needed. With no federals before him, he slipped away to reinforce P. G. T. Beauregard's army at Bull Run and play a key role in the rout of McDowell on July 21. By this time, Patterson's chief concern was not the Confederates but rather the disintegration of his own army as the ninety day terms of service of many of his volunteer regiments began expiring. Officers and men stacked their weapons and headed home. He felt he had no choice but to abandon Charlestown on July 21 and backtrack to Harper's Ferry. Patterson, now shouldering much of the blame for the disaster at Bull Run, was relieved of his command by Major General Nathaniel Banks and mustered out effective July 27.[18]

Nathaniel Banks, the forty-five-year-old former governor of Massachusetts and Speaker of the House of Representatives was one of the army's highest-ranking officers. Appointed a major general of volunteers by Lincoln in the spring of 1861, despite almost no military experience, he looked his rank. Perfect posture, piercing gray eyes, and a shock of hair that swept over his forehead gave Banks an appearance of determination. And he wore his uniform well. From his kepi, cocked at just the right angle, to his polished boots and spurs, he exuded the "air . . . of one used to command."[19]

At Harper's Ferry, Banks reorganized what was left of the Army of the Department of Pennsylvania into the Department of the Shenandoah. Stone's command remained pretty much intact, becoming the 3rd Brigade. On July 28, Banks moved the bulk of his men across the Potomac and back into Maryland, where they remained. Eventually the department would become Bank's division of the Army of the Potomac, but by that time Stone had left for a more prestigious assignment. On August 3, in anticipation of his promotion to brigadier general, he was relieved of his duties in the Department of the Shenandoah and directed to report to Major General George B. McClellan. McClellan, who had achieved some success in western Virginia, had been summoned to Washington to rebuild McDowell's shattered army, and sought to surround himself with promising officers such as Stone.[20]

For his part, Stone was pleased to leave Banks. As with other West Pointers, he disliked serving under a major general who owed his position to political ties rather than military training and experience. McClellan, however, was another matter. Stone knew of his sterling West Point record and had likely known him there. He most certainly had become acquainted with his new commander during the campaign for Mexico City and had assisted him with his railroad route exploration in the 1850s. McClellan was every inch a professional soldier, and someone Stone believed he could trust to make sound decisions and offer inspiring leadership.

By early August, Stone had at last obtained his commission as a brigadier general of volunteers, backdated to May 17. Now General Stone in rank as well as responsibility, he was thirty-six, one of the younger generals, and a rising star in the rapidly growing Union army. In the meantime, the army had created a brigade known grandly as the Corps of Observation. Originally Scott wanted General Charles Sanford to lead the new brigade, but ill health and the impending expiration of his term of service precluded him from accepting the assignment. Aware of Stone's ability and knowledge of the region, Sanford recommended him for the job. Scott then inquired of McClellan "whether or not it would be the wisest course to put Brig. Genl. Stone in command of the Corps."[21] McClellan viewed the general-in-chief's remark as more an order than a question and appointed Stone to the position.

As the name implied, the main purpose of the Corps of Observation was to patrol the Potomac above Washington to Point of Rocks and keep an eye on Confederate activities. By this time, Stone had gained a reputation as the army's expert on that stretch of the river. A nervous McClellan also expected the Confederates to try and invade Maryland. The corps was to be the tripwire, using reinforcements from Banks on the right to help counter any rebel move. Stone and Banks would ensure that "the enemies passage of the river & subsequent advance should be opposed and retarded" until McClellan and the bulk of the Army of the Potomac could come to their assistance. Finally, McClellan gave Stone leeway to attack small enemy concentrations across the river if such actions could be taken with little or no risk to his own men.[22]

Following a brief but welcome reunion with Maria Louisa and Hettie in Washington, Stone marched his command back through Rockville and onto Poolesville, where he once again established his headquarters. From there his men fanned out along the Potomac, occupying high ground from which they could observe enemy movements. He also sent reinforcements to sentinels at Edward's Ferry, Monocacy, and other crossings. Although he did not attempt any attacks into Virginia, Stone did suggest that if he were provided with long-range artillery he could "stir up the intrenchments [*sic*] erected for the defense of Leesburg, and perhaps make them betray the power of their guns, if they have any in position, which I doubt."[23] McClellan did not respond, but throughout the summer and into the fall, he did assign more regiments to the corps. By October, it had attained the status of a division.

Stone's division (still called the Corps of Observation) consisted of three infantry brigades. The first, under Mexican War veteran, former congressman, and former territorial governor of Minnesota brigadier general Willis Gorman, contained the 2nd New York State Militia, 1st Minnesota Infantry, 15th Massachusetts Infantry, and the 34th and 42nd (Tammany) New York Infantry regiments. The second brigade, commanded by civil engineer, transcontinental explorer, and McClellan's former aide-de-camp, Brigadier General Frederick W. Lander, was made up of the 19th and 20th Massachusetts infantries, the 7th Michigan Infantry, and a company of Massachusetts sharpshooters. The third brigade—also known as the California Brigade, although it contained almost no

Californians—was led by Colonel Edward D. Baker. Baker was a Mexican War veteran, sitting United States senator from Oregon, and a close friend of Lincoln. His brigade included the 1st, 2nd, 3rd, and 4th California infantries,which had been recruited largely in Pennsylvania and New York. A smattering of other units, including six companies of the 3rd New York Cavalry, some mounted troopers from Maryland known as Putnam's Rangers, and three artillery batteries rounded out the division.[24]

As his new command assembled, Stone realized that most of the men were raw recruits with precious little military training or knowledge. They were not used to obeying orders unconditionally and had not yet coalesced into a functioning division. Many were away from their homes for the first time in their lives and fell victim to the temptation to loot and abuse local civilians—the very folks from whom the army was trying to secure loyalty and support. The military establishments of both sides considered such actions to be beyond the accepted rules of warfare, and they ran counter to Stone's personal sense of honor and morality. The general, known as an excellent disciplinarian, therefore devoted a considerable amount of effort into turning this eclectic collection of Boston Brahmins, Ivy League scholars, farmboys, factory hands, store clerks, fishermen, and whalers into soldiers equal in proficiency and deportment to the regulars.

To the young, Harvard-educated men of the 20th Massachusetts, Stone seemed in many ways a model general. His posture remained straight as an arrow, and although his hairline had receded a bit, accentuating his prominent forehead, he wore his locks long in the back—hanging just over his collar. He sported a moustache and goatee in the style of Napoleon III, popular among military men of the day. He was neither a blowhard nor a bully. When he spoke, his voice was soft but deliberate, often pausing to ensure that his meaning was clear. Officers admired his military bearing, organizational skills, attention to detail, and impeccable manners (even his opponent, Colonel Eppa Hunton, commanding the 8th Virginia Infantry at Leesburg, remarked that Stone "was a gentleman and conducted war in a gentlemanly manner"),[25] but he lacked the congeniality and comradery the volunteers expected from their commanders. He was not someone that they could warm up to. To most he came across as stiff and standoffish. Although he had demonstrated an ability to motivate his men, as with the District of Columbia Militia, by the fall of 1861 Stone had become a general who relied increasingly on written orders rather than personal leadership. This is not to say that he spent all of his time behind a desk. He often could be found inspecting sentinels and picket posts. The historian of the 20th Massachusetts recalled, "No officer on guard would dare sleep or neglect his duty for General Stone might appear at any moment."[26] Such surprise visits, however, served more to intimidate than inspire.

The status of people of color in the area under his command also alienated Stone from many in his command. Although President Lincoln had been careful to sidestep the issue of slavery in his call for volunteers to restore the Union, a number of the new recruits, particularly the Massachusetts men, saw the war as an opportunity to strike a blow at the odious institution. In particular, they believed that runaway slaves who managed to reach their camps should be protected and not returned to their owners. The Union high command, including Scott, McClellan, and Stone, on the other hand, did not view the conflict as a crusade against slavery. They adhered to the Fugitive Slave Act of 1850, which held slaves to be the property of their owner and returned to them. They further prohibited soldiers from encouraging slaves to be insubordinate to their masters. Stone insisted that his troops observe the laws of the state of Maryland (a slave state), and on September 23 issued an order to the Corps of Observation forbidding any interference with slavery. This directive did not win Stone many supporters among the abolitionists in his division, and would later be used against him by his enemies.[27]

The issue of slavery notwithstanding, by mid-October Stone felt his division had improved to the point that it was ready for more action than just watching the enemy. He had become convinced that the rebels at Leesburg were in a defensive posture, showing no indication that they wished to cross the Potomac in force. He communicated this belief to McClellan, who, having received other reports that the Confederates at Leesburg would likely withdraw if provoked, saw an opportunity to make an offensive gesture with little risk. For some time politicians in Washington with short memories had been pressuring him to move against the Confederates. Once again they raised the cry, "On to

Richmond!" But the Union commander was not yet ready to commit his new army. Leesburg offered a chance to appear to be aggressive without exposing his troops to a major battle. Therefore, he had his adjutant telegraph Stone at eleven o'clock on the morning of October 20 directing him to observe Leesburg to see if a reconnaissance in force by a division under General George A. McCall, who had occupied Dranesville, Virginia, just ten miles away, would prompt the Confederates to evacuate. The order concluded with the suggestion that "perhaps a slight demonstration on your part would have the effect to move them."[28] The message may have seemed ambiguous, but Stone took even a hint from his commanding general as an order. Combined with earlier statements from McClellan giving him broad discretion in the deployment and movement of his division, it persuaded him that he finally had his superior's approval to do more than just observe.

Stone wasted no time in acting upon McClellan's suggestion. For the past four months he had to content himself with marching about western Virginia under Patterson's dilatory leadership and looking across the Potomac at the enemy. During that time he had been refining plans to take Leesburg, and now McClellan had given him the opportunity to do just that. On the afternoon of October 20, he ordered Lander's and Gorman's brigades into positions along the Potomac at Edward's Ferry and Conrad's Ferry, and on Harrison's Island in the middle of the river opposite the highlands known as Ball's Bluff. They even made a pretense of loading into boats, which could be seen clearly by the Confederates. He lobbed a few artillery rounds, "to produce the impression . . . that a crossing was to be made," but the shelling only had the effect of scattering some rebel pickets. Then Stone sent a few boatloads of men over and back, but none of these provocations drew any serious response.[29]

This was curious. Stone had made every attempt to convince the rebels that an attack was imminent, yet they made no moves to counter it or retreat. He then dispatched a scouting party across the river from Harrison's Island to try and find out the location and strength of the enemy. The party's leader, a novice captain from the 15th Massachusetts by the name of Chase Philbrick, however, did not receive his orders until late afternoon on October 20 and did not reach the Virginia side until dusk. In the meantime, Stone ordered Lander's and Gorman's brigades to return to their original positions at sunset, ending the demonstration, and settled down to await the scouting party's return.[30]

Philbrick made it back to Harrison's Island by ten P.M. and reported to his regimental commander, Colonel Charles Devens, that he had found a little-used path up the bluff and travelled to within a couple of miles of Leesburg. He had discovered a rebel camp of about thirty tents between the town and the river, but was unable to detect any campfires, movement, or pickets. Devens in turn sent his quartermaster, Lieutenant Howe, with this information to Stone at his field headquarters at Edward's Ferry. Combined with intelligence he had already received that the Confederates were likely to withdraw from Leesburg, and that McCall was still in front of Dranesville and would support him, Philbrick's report prompted Stone to take a "very nice little military chance." He made his fateful decision just before midnight. He would turn McClellan's "slight demonstration" into a raid on the seemingly unprotected camp and, perhaps, Leesburg itself.[31]

TRIPPED UP
BY CIRCUMSTANCES

S TONE was anxious to begin the raid on the rebel tents and possible advance on Leesburg. From his Edward's Ferry headquarters he issued orders to Colonel Devens to take five companies of the 15th Massachusetts from Harrison's Island by boat across the Potomac to the heights known as Ball's Bluff, scale them, and then conceal his force until daybreak when he would attack and destroy the camp. Devens's movement would be accompanied by a company of the 20th Massachusetts under its commander, Colonel William Lee, who had attended but not graduated from West Point. This company would cross the river, climb the bluff, and be prepared to cover Devens should he need to retreat. After this, Stone's instructions became less clear. If he encountered little or no opposition, Devens, after conducting a reconnaissance of the area, could hold his position and report back to Stone. Or he had the option of withdrawing to Harrison's Island. It was up to the colonel's own discretion. Stone, however, obviously favored the former course. He believed that the Confederates had committed "one of those pieces of carelessness" he could not help but exploit.[1]

At the same time that Devens was supposed to be crossing into Virginia, Stone dispatched two companies of infantry and some cavalry from Gorman's brigade across the river at Edward's Ferry on a reconnaissance mission and to serve as a possible distraction from the activity at Ball's Bluff. He also sent five companies of the 15th Massachusetts to Smart's Mill, about a mile upstream from Ball's Bluff. Here was a ford that Stone had personally reconnoitered and determined to be crossable with little or no use of boats. The Massachusetts men were to occupy and hold the mill so that Devens could use the location should he need to make a quick retreat. And, if fighting did develop, Stone could rely on McCall's troops, which he still believed to be in front of Dranesville. It seemed a simple, safe plan.[2]

Almost at once, however, things began to go awry. Although Stone issued his orders between ten and eleven P.M. on October 20, Devens and Lee did not receive them until midnight, delaying the start of the mission. Devens was not enthusiastic about leading upwards of three hundred untested soldiers on a night raid into enemy territory with only sketchy intelligence. He took his time in arranging his troops for the crossing. The Potomac was also running high, with a swift current, and Devens could locate only three boats, each capable of carrying twenty-five to thirty men. In addition, Lee deviated from Stone's plan by taking an additional company over to support Devens. As a result, it took about five hours to ferry the 15th and 20th Massachusetts companies over to Virginia, where they found the path up the bluff to be steep, narrow, and choked with brush. Once on the heights, Philbrick led the raiding party across a meadow studded with oak and locust trees and into a wooded area. There the boys from Massachusetts would wait for the sunrise and their first taste of combat.[3]

"At the first symptom of light," Devens moved out of the woods into an open field to look for the rebel camp. But there was no camp. In fact, there were no rebels anywhere, only a line of trees (some accounts said rows of haystacks), which in the misty moonlight must have looked like tents to the nervous, inexperienced Philbrick. Now, what to do? Stone had given him the option to withdraw, and without a target to attack, that might have been the prudent move. Had he done so, there would have been no battle at Ball's Bluff. But Devens was unsure. There did not seem to be any immediate danger of discovery, so he chose to keep his men secreted in the woods and send Howe to report to Stone and receive further instructions. Stone was likely disappointed about the camp. Since Devens and Lee were in Virginia and had not yet been discovered, however, he sent Howe back with instructions to turn the operation into a reconnaissance. Devens complied and began advancing tentatively toward Leesburg.[4]

The Confederates, meanwhile, soon learned of the Union presence in Virginia. Pickets from the 17th Mississippi detected suspicious movement near Ball's Bluff and exchanged a few shots with an unseen foe. About the same time as Devens was discovering Philbrick's error about the camp, Colonel Nathan Evans, commanding the rebel brigade defending the Leesburg area, received word of this skirmish, as well as an encounter with some Yankees opposite Edward's Ferry. Evans initially was cautious. With multiple reports of federal activity in the area he needed a better understanding of just what was going on and decided to wait and see how the situation developed. He did not have long to wait.

Back at Ball's Bluff, Devens became involved in a "sharp fight" with a company of the 17th Mississippi, which had maneuvered so as to block the way to Leesburg. Although they outnumbered their opponents, the Massachusetts men had been surprised, and many carried smoothbore muskets inferior in both accuracy and range to the rifles of the Mississippians. After suffering one killed and a number wounded or captured, Devens fell back to the woods and then to a ridge above the bluff. The rebels knew better than to pursue a superior foe too far, and their reinforcements had not yet arrived. Devens, therefore, still had the opportunity to retreat to Harrison's Island, ending the fight then and there. Instead he chose to remain in Virginia, again sending Howe to Stone to ask for support or orders to withdraw. Before leaving the bluff, Howe encountered Lee, who added an ominous caveat: "Tell Stone that if he wants to open a campaign in Virginia, now is the time."[5]

On his way to Edward's Ferry, Howe ran into the five companies of the 15th Massachusetts preparing to occupy Smart's Mill. When he told them of their comrades' difficulty over in Virginia, the officer in charge, Lieutenant Colonel Ward, chose to deviate from Stone's instructions, cross the Potomac at Ball's Bluff, and go to the assistance of the regiment. While this added to the number of troops available to Devens, it removed his easy escape route and increased the strain on what few boats he had. In addition to the loss of Smart's Mill, Stone's plan received another blow when McClellan became convinced that the double feint of Stone and McCall had failed to prompt the Confederates to move. Believing that Stone would not attempt to cross the Potomac in force and that the Corps of Observation was safe in Maryland, he recalled McCall's division from Dranesville. It would not be available to provide assistance to the operation at Ball's Bluff. McClellan, however, chose not to communicate this information to Stone. Then Colonel Baker arrived at Edward's Ferry.[6]

At the same time that he had ordered Devens across the Potomac, Stone ordered Baker, who was then camped some five miles away near the mouth of the Monocacy River, to move the 1st California down to Conrad's Ferry as soon as possible and, after feeding them a good breakfast, prepare the rest of his brigade to march. Baker, however, had only just over half of the regiment on hand, the remainder being on picket duty farther upstream and unavailable. Nevertheless, by 8:30 A.M. Baker and his second-in-command, Lieutenant Colonel Isaac Wistar, had positioned the 1st California on the Maryland shore at Conrad's Ferry, ready to cross when ordered. When the order did not come, an impatient Baker then mounted his horse and galloped off to meet with Stone and find out just what he was supposed to do. He arrived at Edward's Ferry at about nine A.M.[7]

The meeting between Stone and Baker forms the crux of the controversy surrounding the Battle of Ball's Bluff. Although a professional soldier and in command, Stone may have been a bit intimidated by his distinguished subordinate. Baker was thirteen years Stone's senior, and had been a congressman when Stone was still at West Point. During the Mexican War, Baker was a colonel and led a volunteer regiment while Stone was a lieutenant in charge of a single siege battery. After the war, Baker's star continued to rise as he became a US Senator and influential leader in the young Republican Party. Since resigning from the army in 1856, Stone, on the other hand, had encountered one difficulty after another. It was well known that Baker had been offered a brigadier's commission by President Lincoln, although he had turned it down in order to continue serving in the Senate. In

September, however, he had been tendered a commission as a major general, and was still considering this offer. He could, quite possibly, become Stone's superior officer. And Baker was a close personal friend and political ally of the president. While he did not defer to the celebrity colonel, Stone likely did not use his command voice, which his detractors described as curt and "imperious," in conversation with him. He had no reason to distrust Baker's judgment, but he knew the colonel possessed no formal military training, and that his experience in Mexico, while valiant, had been limited. It would be necessary to provide him with a thorough understanding of the situation and clear orders.

Stone talked with Baker for about a half-hour at Edward's Ferry. He discussed how Devens's mission had changed from raid to reconnaissance. He also explained that he had sent a small diversionary force from Gorman's brigade across the river, and that McCall's division was ready to provide assistance if needed. He mentioned what he knew about the availability and capacity of the boats at Harrison's Island, but assumed that Baker would sort that out for himself. Stone then instructed Baker to proceed to Ball's Bluff and take command of the operation. In particular, he wanted to get reliable information on rebel positions and movements. He gave him the option either to withdraw the troops already in Virginia back to Harrison's Island or to cross additional units and explore, "as far as it was safe," toward Leesburg. Baker was to use his own discretion in making this decision. If Baker chose to advance, he would have Devens's and Lee's men, his own 1st California, and the 42nd New York, as well as some artillery, at his disposal. At the time of their meeting, neither man knew that fighting had broken out over in Virginia, but Stone ordered that if Baker did encounter any opposition he was not to press forward unless the enemy force proved clearly inferior to his own. Above all, he did not want the colonel to be lured into a trap.[8]

It is not known whether Baker made any comments or questioned Stone about his orders other than to ask if he had the entire command at his disposal. Stone replied in the affirmative, but his is the only account of the meeting. Before departing, however, Baker asked for some written authority for assuming command. This may have irritated the general, since he felt he had provided thorough and clear verbal instructions. Nevertheless, Stone quickly had a brief order written out and handed it to him. He directed Baker to assume command, and, once again, stated that if he found fighting going on "in front of Harrison's Island," he was either to bring his 1st California across or retire, "at your discretion," the regiments of Devens and Lee. According to Stone, the colonel seemed pleased to receive the order. He stuck the paper in his hat, mounted his horse, and galloped away.[9]

Believing the situation at Ball's Bluff to be in competent hands, Stone telegraphed McClellan at 9:45 A.M. that he had ordered reconnaissances into Virginia. Devens was across the river with five companies and men from Gorman's brigade held the shore opposite Edward's Ferry. He did not mention Baker. This was the first the commanding general knew of federal troops crossing the Potomac in substantial numbers. He had not ordered a movement into Virginia, but trusted that Stone knew what he was doing. He telegraphed back, "I congratulate your command. Keep me constantly informed." Pleased with McClellan's reply, Stone turned his attention to new information brought by the peripatetic Lieutenant Howe.[10]

On his way to Harrison's Island, Baker had encountered Howe, who was heading to Stone's headquarters with the report on the clash between Devens and the 17th Mississippi. Upon hearing this news, the colonel made a snap decision. He would ride to the sound of the guns. With an oratorical flourish he proclaimed to the weary lieutenant that he would immediately lead the men available to him over to Virginia. This was before he had even gotten close to the scene of the fighting and had a chance to evaluate the rebels' strength and position. Before parting, Howe also informed Baker that Smart's Mill had not been occupied, contrary to Stone's orders, but this did not faze him. In the colonel's mind there should be no need for retreat. Besides, the extra five companies of the 15th Massachusetts ought to be helpful in routing the Confederates. That there would be a pitched battle at Ball's Bluff was now a foregone conclusion.[11]

Despite his bold words, Baker did not seem to be in a hurry to get over to Ball's Bluff. He spent well over an hour on the Maryland side superintending the hoisting of a boat out of the Chesapeake and Ohio Canal and placing it in the Potomac. While this vessel would prove useful in supplementing the transportation available at Harrison's Island, overseeing its launching into the river may not have been the best use of the colonel's time. Stone later commented that this was a job better suited to a junior officer. When Howe, returning to Virginia, again encountered Baker, he asked him if he had any orders. The colonel replied that he should tell Devens his men had performed "nobly," but gave no further instructions. For now, Devens was on his own.[12]

Howe's report surprised and probably disappointed Stone. The reconnaissance had been discovered, and the 15th Massachusetts had suffered casualties and been pushed back. On the other hand, he knew that Howe had told Baker about the action, and that Baker was going to cross his troops to challenge the rebels. Stone expressed full confidence in Baker. He again telegraphed McClellan at 11:10 A.M. to say that his troops had engaged the enemy across from Harrison's Island. Although he provided no details about his orders to Baker or anything else going on in Virginia, he did add an upbeat note: "Our men are behaving admirably."[13]

Stone's message definitely got the commanding general's attention. McClellan had not anticipated, nor did he really want, any type of battle. Nevertheless, since fighting had already broken out, he saw the possibility of gaining an advantage. If it could be done fairly bloodlessly, a victory might squelch the criticism of his leadership coming out of Congress. Anticipating that McCall's division might be needed, he telegraphed him to remain in Dranesville, but by the time he received the message the general had already returned to his camp some twenty miles away. In his reply to Stone, after inquiring about the enemy's strength and telling him that he could call on Banks for reinforcements, McClellan stated that he might have him take Leesburg that day and asked what force such a movement would require.[14]

Stone interpreted his commander's telegram as a sort of permission to have Baker advance on Leesburg, but from what Devens had relayed he estimated that the colonel would be facing a strong enemy. Assuming that Baker already had taken command across the Potomac, at 11:50 A.M. he sent him a warning that there were upwards of four thousand Confederates opposing him (a gross overestimate), but keeping open the option of attempting to push the rebels back. Stone also telegraphed McClellan with this information along with the optimistic assessment that his men could take Leesburg by the end of the day. He concluded, however, with the first hint that things might not be going quite according to planned: "We are a little short of boats."[15]

Unbeknownst to Stone, when Baker got around to replying to his general's earlier message at 1:30 P.M., he was still on Harrison's Island. His communication, though, gave no hint that he had not yet made it to the scene of the fighting. Rather, it let his general know confidently that he would be bringing his entire brigade into the fray.[16]

Now fully expecting Baker to move on Leesburg, Stone turned his attention to his left. There he had been orchestrating a leisurely crossing of additional units from Gorman's brigade at Edward's Ferry. His plan was to distract the rebels, drawing troops away from the main attack. When Baker began to push the enemy back, he would attack their flank. But while everything appeared to be going well, Stone still did not know: (1) the five companies he had expected to cover a possible retreat at Smart's Mill had instead joined the rest of their regiment at Ball's Bluff; (2) a unit of cavalry that he had dispatched to the Virginia side to act as scouts for Devens simply stood around for a while on the shore and then returned without ever reaching the scene of the action; (3) as of 1:30, Baker had not yet made it to the battlefield and taken command; (4) Colonel Evans had determined the crossing at Edward's Ferry to be a feint, and so decided to concentrate his men against the main force at Ball's Bluff; and (5) McCall was not advancing from Dranesville. Instead, McClellan had pulled him back to Langley.

One of the most serious problems faced by Stone and other Civil War commanders was communication. The army relied on three principal means of transmitting information. The telegraph provided almost instant contact, but was limited to locations where wire was strung and operators were available. In the case of the battle at Ball's Bluff, the nearest telegraph station was four miles from Edward's Ferry. McClellan had set up a system of semaphore flag stations connecting Stone with McCall and Banks, but these were of little use at night, and accuracy depended on the skill and training of the signalmen. Finally, couriers carried messages, but they could be slow and ran the risk of getting lost or being intercepted.

As a result, Stone knew that he was operating in the dark regarding much of what had been occurring over in Virginia and at headquarters. He would later state, "I was obliged to proceed very much on my own ideas of what was taking place elsewhere."[17]

At two P.M. Stone telegraphed a message to McClellan based as much on wishful thinking as hard information. There was "sharp firing" over in Virginia, but Baker appeared to be moving forward, and the skirmishers from Gorman's brigade had advanced a mile toward the enemy in preparation for turning his right flank. A short time later, he supplemented this with descriptions of the boat situation and the roads to Leesburg. Around four P.M. McClellan's response arrived in the form of a coded message. He reiterated that Stone should rely on Banks for more troops, and told him to "take Leesburg." He also instructed that future messages be sent using a cypher. This frustrated Stone, who did not have the cypher key. He therefore did not receive the order regarding Leesburg. Instead, he dispatched a cryptic reply that, while he had received the box, he had no key. Headquarters staff took this literally and launched a wild goose chase for a physical key before realizing Stone's meaning. McClellan was beginning to be uneasy.[18]

In the meantime, Stone had sent a short follow-up message to McClellan and Banks. Almost all of his troops were across the Potomac, Baker on the right and Gorman on the left (the latter somewhat of an exaggeration, since not all of Gorman's brigade was in Virginia), and Baker was "sharply engaged." In a subsequent message to Banks, he requested a brigade be sent down, not because his own men were in trouble but to act as a reserve, should it be needed for the push on Leesburg. It was the last good news either general would receive from Stone.[19]

By four in the afternoon, Baker had been at Ball's Bluff for less than two hours. Upon his arrival, he had congratulated Lee on the prospect of a battle and cheered the men of the 20th Massachusetts by promising them a chance to fight. He relieved Devens (who was relieved to be relieved), inspected the line, and began deploying units of the 1st California that had started to arrive. He asked both Lee and Wistar about his troop dispositions, and, while neither forced the issue, both told him the battle would be won or lost on the Union left. Baker ignored this advice, even going so far as to tell Wistar, "I throw the entire responsibility for the left wing upon you. I throw it upon you. Do as you like."[20] When Milton Cogswell appeared leading the first of the 42nd New York, Baker asked him about his troop placements as well. The West Pointer, who had taught infantry tactics at the academy, minced no words. He found Baker's positions to be "very defective," stressing that the wooded hills on the left commanded most of the battlefield and that they ought to be secured before the Confederates took them. Despite these recommendations from two of his regimental commanders and his own second, Baker chose not to alter his battle plan. He would keep a relatively strong point in the woods on his right, while on the left his men would be strung out in a weaker formation in the open along the cleared area. As a result, the rebels were able to occupy the high ground and rain deadly fire on the Union line.[21]

Baker now became cautious. He sent out two California companies to scout the left, but these were turned back. After beating off some initial rebel assaults, he realized the danger on his left and started shifting troops in that direction. The Confederates, however, had reinforced their position and began pressuring the Union left and center. As the fighting intensified, Baker dismounted and walked calmly about the battlefield, sword in hand. He offered consoling words to the wounded and rallied those remaining to fight on. His thoughts may have wandered back to Cerro Gordo, where his volunteers swept the field. But now he was facing seasoned veterans of Bull Run, not Santa Anna's demoralized conscripts. The rebels were motivated, well led, and well armed. Lee, Wistar, and Cogswell may have been right all along. He should have concentrated on the left. Although the day was not necessarily lost, there would be no glorious advance on Leesburg. His men shouted for him to take cover, but he ignored their pleas. A commander who led by example must seem impervious to danger. Baker turned to an aide and ordered him to go to Stone and request reinforcements. A moment later, sometime between 4:30 and 5:00 P.M., Baker was shot and killed near the center of the Union line. No one knows exactly when or by whom.[22]

Although Baker's death demoralized the Union forces, it by no means ended the battle. Those companies still engaged continued to fight on. Lee assumed temporary command. Devens, who had no desire to be back in charge, offered no opposition. Lee made a pessimistic appraisal of the situation. The Confederates were pressing even harder, and the best alternative seemed to be to get as quickly as possible back to Harrison's Island. Before he could effect the withdrawal, however, Cogswell arrived to challenge his authority. The three colonels met briefly, and Lee and Devens deferred to the army regular. Cogswell was not ready to give up. He organized a desperate attack on the Confederate right in hopes of breaking through to Gorman, or, possibly McCall, who was still believed to be marching on Leesburg. But by the time he completed the arrangements for the charge it was dusk. His soldiers were exhausted, and the rebels sensed victory. While the California regiment joined the Tamany in the assault, the Massachusetts men, most of whom had been in the field since midnight, were held back due to a mix-up in orders. The breakout failed, and Cogswell had no choice but to order a retreat. Devens, who had regained some confidence, demurred and now wished to make a fight of it. He asked Cogswell to repeat the order in the presence of his major. He did, Devens obeyed, and the withdrawal from the bluff was on.[23]

What followed was more of a rout than a retreat. By now, many of the Union troops did not need to be told to pull back. Lines broke and men made their way as best they could to the edge of the bluff. In the last light of day they ran, rolled, and slid down to the river. There chaos ensued. They were, indeed, "a little short of boats" to get back to Harrison's Island. Seeing this, a small group headed upstream along the riverbank to cross at the undefended Smart's Mill. Most others panicked. As rebel bullets rained down, they clambered into what boats they could find, including vessels being used to transport the wounded. Some boats swamped, spilling their contents into the cold, fast Potomac. Officers ordered soldiers to throw away their weapons and swim for it. Few knew how to swim, but, faced with either being shot or captured, tried it anyway. Even strong swimmers found it difficult to battle the river's strong current. A number drowned, their bodies being swept downstream toward Washington. Those who remained on the Virginia side became prisoners, including Cogswell and Lee.[24]

Stone had received nothing in writing from Baker after his 1:30 message, but the two men kept in contact via the verbal reports of officers who shuttled back and forth between the battlefield and

Edward's Ferry. At 2:30 and 3:00, Baker let Stone know that he was holding his own against strong rebel opposition, but hinted that he might not be able to advance on Leesburg. In his 3:45 reply, the general urged Baker to maintain his position, but relieved him of the responsibility of moving forward if he decided he could not. In a bit of micromanagement, Stone also admonished the colonel not to let his men get too fatigued or hungry. Although likely disappointed by the possibility that Baker might not be able to take Leesburg straightaway, Stone was nonetheless still confident that the battle was progressing satisfactorily. Assuming that the colonel would remain on the Bluff, the impending arrival of McCall from Dranesville and Gorman's advance would crush the Confederate right and free Baker to break out. With Banks's brigade in reserve, the day would be won.[25]

Baker never received Stone's 3:45 message. By the time the courier reached Harrison's Island, he learned the colonel had been killed. He doubled back to tell Stone the "melancholy tidings," arriving at Edward's Ferry at 6:00. The general was shocked, but not panicked. He telegraphed McClellan, "Colonel Baker has been killed at the head of his brigade." He did not know it at the time, but this terse message would reach ears far more important than those of his commander.[26]

Earlier in the day Lincoln had received word that Baker was involved in a battle, and worried that his friend's impetuosity might get him into trouble. Rather than remaining in the White House, he went to McClellan's headquarters to await developments. The two men were talking about the action in Virginia when the telegram arrived regarding Baker's death. The commanding general first put the paper in his pocket without showing it to the president. A short time later while talking with reporters, Lincoln was escorted into the telegraph office to be given the bad news. A reporter present noted that "the death of his beloved Baker smote upon him like a whirlwind from the desert." The chief executive was visibly shaken, "his face pale and wan, his heart heaving with emotion." Just the day before, he and Baker, along with young Willie, had spent the afternoon strolling the White House grounds and relaxing under the trees. Mrs. Lincoln had given him a bouquet as he left. Now his old friend—probably his closest friend, for whom he had named his second son, Edward

Baker Lincoln—was dead. A weeping president found it difficult to maintain his footing when leaving the building.[27]

After telegraphing McClellan and Banks the news about Baker, Stone rode toward Ball's Bluff. He would review the situation and take personal command, if that was required. Although concerned that his field commander had been killed, and suspecting "some trouble on the right," he still did not know the extent of the disaster.[28] He assumed one of the other colonels present must have taken charge and the troops were still engaged. Reinforcements could be expected from Banks and McCall. Baker's loss had been a blow, but not necessarily a fatal one.

The scope of the defeat became clearer as Stone approached the landing opposite Harrison's Island. He glimpsed Baker's body being transported from the scene and encountered a steady stream of cold, wet, and demoralized men moving up from the riverbank. Many had lost their muskets. He learned from these survivors of the rout on the bluff and the chaos of the retreat across the river. Stone later recalled, "I met more and more men and began to fear we had a disaster."[29]

Stone had to admit the battle at Ball's Bluff had been lost. As the last Union stragglers made their way to Harrison's Island there was nothing to do but dig in against a possible Confederate counterattack. While surveying conditions opposite the bluff and beginning the reforming of companies and regiments, he realized that the troops from Gorman's brigade on the Virginia side opposite Edward's Ferry might now be in danger. He still believed the enemy numbered about four thousand, and, having dispatched the Yankees at Ball's Bluff, they might turn on the other invaders. After placing the defeated force under the command of Colonel Edward Hinks of the 19th Massachusetts with instructions to deploy pickets to detect any Confederate attempt to cross the Potomac, and to hold Harrison's Island, Stone rode back to Edward's Ferry.[30]

Stone's first inclination was to withdraw Gorman's men back to Maryland. He had lost much of one brigade and did not intend to lose another. At 9:30, he telegraphed McClellan that he was "occupied in preventing further disaster, and try to get into a position to redeem." He then summarized the day's sad events while trying to put the best spin possible on the situation:

We have lost some of our best commanders—Baker dead, Cogswell a prisoner or secreted. . . . All was reported going well up to Baker's death, but in the confusion following that, the right wing was outflanked. In a few hours, I shall, unless a night attack is made, be in the same position as last night, save the loss of many good men.[31]

Stone still believed that McCall's troops might advance from Dranesville. He advised McClellan that any such movement must be done cautiously, and stated that he was withdrawing Gorman's men from Virginia.

McClellan later would comment that Stone's dispatch did not give him a complete understanding of what had happened. He still believed most of the Corps of Observation to be across the Potomac. As a result, at ten P.M. he directed Stone to hold on in Virginia, "at all hazards," until reinforcements from Banks could arrive. At that time he could either remain or pull back to Maryland. Stone must have been perplexed by this response. He no longer had troops in Virginia on his right, and was already preparing to defend Harrison's Island. Sensing that his commander might be a bit confused about the situation, at eleven o'clock he sent a more complete description of his position:

We hold the ground a half mile back of Edward's Ferry on Virginia shore. Harrison's island has parts of thirteen companies, only seven hundred (700) men and will soon be reinforced by one hundred fresh men, besides what Hamilton [Banks's brigade commander] brings. I cover the shore opposite with guns, and am disposing others to help the defence of Harrison's. I think the men will fight well. Entrenchments ordered this morning.[32]

Stone also countermanded his order to begin withdrawing Gorman's brigade from Virginia. Instead he sent more men along with entrenching tools and instructions to dig in.

In addition to his communications with McClellan, Stone had been in touch with the president. Returning to the White House after receiving the news of Baker's death, Lincoln dispatched a messenger from Poolesville to get information from Stone about the battle. In a rambling telegram, written by a man too long without sleep and under stress, the general stated that he could not explain what had gone wrong, "Our troops under Col. Baker were reported in good condition and position within 15 minutes of the death of Col. Baker." He then proceeded to detail the losses that had been incurred, down to the number of cannon captured. Nowhere did Stone express any remorse or condolence over the death of the president's friend. What Lincoln thought of this message was not recorded, but it did not endear the general to him.[33]

Finally, at 11:30 on the night of the twenty-first, Stone telegraphed McClellan urging that an advance be made from Dranesville. McClellan responded, rather offhandedly, that this was not possible and that he had to rely on troops from Banks. At last, Stone learned that McCall was not coming.[34]

The first reinforcements from Banks arrived early on the morning of October 22. By the time Banks himself appeared at four A.M., his men had already begun crossing the Potomac to reinforce Harrison's Island and Gorman. Although McClellan had clearly told Banks to assume command, he did almost nothing. He deferred to Stone, who had been in the field for over twenty-four hours and was by now thoroughly exhausted, to direct the deployment of the brigades at Edward's Ferry and opposite Ball's Bluff.

In reality, Stone, Banks, and McClellan did not have to worry about a Confederate advance on Harrison's Island. The rebels, while elated over their victory, were also worn out and in no condition to launch an attack across the Potomac. And, like the Yankees, they lacked sufficient boats for such a move. If anything, Evans, who knew the enemy had been reinforced, speculated that he now might have to defend against a Union push from Edward's Ferry.

Fighting began opposite Edward's Ferry on the morning of the twenty-second. The 13th Mississippi Infantry, "yelling like demons," surprised pickets of the 1st Minnesota. The ensuing skirmish proved inconclusive. The rebels advanced between three and four hundred yards before federal artillery forced them back. Only one Union and two Confederate soldiers were lost that day, although General Lander, who had come up from Washington, suffered a wound that would prove fatal just over four months later. As a rainstorm approached, the federals held their line and waited for either another attack or orders to advance.[35]

On the afternoon of October 22, McClellan set out to see for himself just what had happened and what needed to be done. By the time he reached Poolesville, he had determined that Harrison's Island could not be defended, and ordered all troops there back to Maryland. The evacuation occurred under the cover of darkness and was accomplished without the loss of a single man.[36]

When McClellan arrived at Edward's Ferry the next day, he found that Banks had not done much to improve the situation besides securing all of the canal boats in the vicinity. Stone, on the other hand, had sent out patrols to determine the Confederate strength and position. These returned without engaging the enemy. This was not surprising because Evans had already begun moving quietly away from Leesburg. Nevertheless, believing the rebels were in vastly superior numbers, McClellan deemed the Union position opposite Edward's Ferry to be untenable and, before leaving, directed all troops to return to Maryland. Stone directly supervised the withdrawal, which was accomplished uneventfully during the night of the twenty-third. By four the following morning, the last boat had departed from Virginia. Stone then had himself rowed across the Potomac and up and down the shore before finally being convinced that "not a man or horse had been left behind."[37]

The Battle of Ball's Bluff was over. Despite wildly inflated estimates at the time, only about 1,700 soldiers on each side participated in the action on the bluff, and over 4,000 crossed at Edward's Ferry. There they faced the 13th Mississippi with close to six hundred men. The Union had lost one brigade commander (and US Senator) killed and two regimental colonels captured. The Confederates lost one colonel. The rebels suffered about forty killed and 117 wounded. Union casualties were more difficult to determine. Just forty-nine were initially listed as killed, with 158 wounded and over 700 taken prisoner. Later calculations put the number of fatalities at 223, 226 wounded, and 503 captured. The severity of the Northern defeat is clear. Although the overall number of casualties seems small—as Stone later put it, "about equal to an unnoticed morning's skirmish on the lines before Petersburg," given the number of participants—the percentage of Union loss at Ball's Bluff far outweighs that suffered at Bull Run and many other larger and better-known battles.[38]

What started as Stone's "nice military chance" had ended in disaster. But who was responsible? Clearly, Stone's subordinates, particularly Baker, bore much of the blame for what went wrong. Baker's ardor outweighed his military ability. His impulsive decision to advance his brigade across the Potomac without first viewing the situation at Ball's Bluff or discussing it with those already involved in the fighting, his failure to secure adequate transportation from and to Harrison's Island, and the offhand dismissal of the advice of his fellow colonels as to the proper deployment of his men, doomed the Union chances. Stone later said of Baker's performance, "He brought [the battle] on, and, I am sorry to say, handled his troops unskillfully in it."[39]

Other Union officers contributed to the debacle as well. Philbrick's misidentification of the Confederate camp prompted Stone to send Devens into Virginia in the first place. Ward's disobedience of orders to hold Smart's Mill deprived the Yankees of an easier and less dangerous route for retreat back into Maryland. And had Devens chosen to withdraw when Philbrick's error was discovered, or once the shooting had started, the battle would likely not have occurred at all.

At headquarters, McClellan quickly denied any culpability for the defeat. His imprecise orders, however, had given Stone the impression that he would support a raid on Leesburg. His failure to communicate regarding McCall's withdrawal from Dranesville also led to Stone's false sense of security that an entire division would be available to assist him. Had Stone known that McCall was not coming, he likely would have been more circumspect in sending his troops across the Potomac.

Stone's other source of reinforcements also proved problematic. Although Banks promptly dispatched badly needed brigades on the double when requested by Stone and instructed by McClellan, once on the scene, the major general, according to McClellan, did little. He allowed Stone to continue issuing the orders and directing the troops until McClellan showed up. Perhaps Banks believed he had been told merely to support Stone or that McClellan was on his way and would soon assume command himself. Whatever the reason, Banks did not do much to improve the situation at Ball's Bluff and Edward's Ferry.

Ultimately, as commander of the Corps of Observation, Stone bore the overall responsibility for the loss at Ball's Bluff. Although he may not have been planning an attack on Leesburg on the night of October 20, Philbrick's incorrect report encouraged him to skew McClellan's suggestion toward a movement more aggressive than just a "demonstration." He may have criticized Baker for wanting to "bring on a battle," but his blood was up as well. The operation that followed, with two fairly widely separated brigades in action, may have just been too complicated for him to handle efficiently. Stone was a talented organizer and administrator, but he did not possess that much combat experience, and had no experience in leading a company, never mind a division, in battle. Given the time it took him to communicate with his officers in the field and with McClellan at army headquarters, he found it almost impossible to keep up with, let alone direct, events in Virginia. He badly misjudged the abilities of his brigade and regimental commanders, only two of whom had formal military training, and, with the exception of Baker, none of whom had ever been under fire. Stone's discretionary orders allowed these men to make the mistakes that cost the Union dearly. General Lander probably said it best when he opined that Stone, "a very efficient, orderly and excellent officer," was, "in the matter of Ball's Bluff . . . tripped up by circumstances."[40]

THE COMMITTEE

A LTHOUGH he had suffered a devastating defeat, Stone remained in command of the Corps of Observation. His immediate responsibilities were to see to the burial of his dead, treatment of the wounded, and, as best as he could, the care of his men who had been taken prisoner. Stone also had to rebuild his shattered division. Rebel activities across the Potomac still needed to be monitored. A feint or raid might be a precursor to an invasion of Maryland. As replacements arrived to augment what remained of the Corps, Stone established more observation posts. He installed viewing platforms and conducted interviews with teamsters and merchants who conducted business across the river, persons caught making unauthorized visits to Virginia, runaway slaves, and Virginians whose sympathies lay with the Union to garner intelligence. In addition, he employed an innovative method of spying on the enemy—observation balloons.

The use of balloons for military purposes had been pioneered by the French in the late eighteenth century and had been used by armies in Europe off and on since then. The United States Army, on the other hand, paid almost no attention to them prior to the Civil War. By the time of the first battle of Bull Run, however,

Professor Thaddeus S. C. Lowe, America's foremost balloon aeronaut, was making valuable aerial observations for General McDowell. The establishment of the civilian Balloon Corps under Lowe soon followed.

In early November, General Gorman spied a rebel balloon over Leesburg. He reported this to Stone, who requested that McClellan assign a similar apparatus to his division. Difficulties in fabricating the hydrogen gas-producing machine delayed the deployment of the balloon *Intrepid* to Edward's Ferry for almost a month. When the craft finally arrived on December 14, Stone accompanied Lowe that day on its first ascent. The experience convinced Stone of the balloon's usefulness in observing movements and fortifications for miles across the Potomac. As often as winter weather and the availability of gas permitted, the *Intrepid* was sent up, with Stone himself going aloft at least one more time. He was able to provide McClellan with the most accurate assessments of Confederate troop strength and locations heretofore available. The potential of aerial reconnaissance so impressed Stone that in mid-January 1862 he requested a second balloon, and when hail and ice damaged the *Intrepid*, he immediately asked for a replacement. Stone might well have kept his head in the clouds and become the army's first aeronaut-general had not events in Washington sent his career plummeting back to earth.[1]

In late December, Stone received notification to travel to Washington and appear before a congressional committee. Unsure of the reason, he telegraphed McClellan asking if he needed to honor the summons. When told that he should, Stone asked that he be directed as to the day he ought to leave his command. Given past events he was concerned that, should anything "unforeseen" happen, it would further damage his reputation if word got out that he and not headquarters had arranged for him to be absent. Something unforeseen would indeed happen to Stone, but not on the banks of the upper Potomac.[2]

Trouble had begun brewing almost before the smoke of battle had cleared from Ball's Bluff. Stone submitted his initial account of the action to McClellan on October 29, after receiving the reports of the principal surviving officers not in rebel custody. Beginning with the "demonstration" on October 20 and ending with Banks's arrival on the morning of the twenty-second, Stone presented a straightforward narration of the battle. While he did not blame anyone directly for the loss, he pointed out the error in Philbrick's scout, the failure of the cavalry detail to carry out its orders, and Ward's unauthorized diversion away from Smart's Mill. He further intimated that Baker may not have handled competently the matter of the boats at Harrison's Island and made a serious mistake in crossing his artillery first, thereby preventing his infantry reinforcements from being present when the Confederates pressed their attack. These were Stone's only initial criticisms of Baker in writing.[3]

Privately, however, Stone may have voiced his disappointment over Baker's performance to McClellan. Three days after the battle, McClellan telegraphed the division commanders of the Army of the Potomac praising the bravery of the men who fought, but stating emphatically: "The disaster was caused by errors committed by the immediate Commander [Baker]—*not* Genl Stone."[4] On October 25, he expanded his criticism of the late colonel in a letter to his wife, Mary Ellen:

> The man *directly* to blame for the affair was Col Baker, who was killed—he was in command, disregarded entirely the instructions he had received from Stone, & violated all military rules and precautions. Instead of meeting the enemy with double their force & a good ferry behind him, he was outnumbered three to one, & had no means of retreat.[5]

Nevertheless, ever mindful of his own reputation, McClellan did take care to note that the battle "was entirely unauthorized by me & I am in no means responsible for it."[6]

Most professional military men agreed with McClellan's assessment of who was responsible for Ball's Bluff. Baker, however, had a legion of supporters among the nation's press and politicians. He had, after all, been a patriot and steadfast supporter both in the Senate and in uniform of the Union and the war to preserve it. Muffled drums sounded as his body was transported to the Congressional Cemetery for temporary internment prior to being shipped to California. Congressmen, senators, cabinet members, justices of the Supreme Court, and hundreds of other mourners joined a weeping President Lincoln in the procession. As the coffin later made its way to Philadelphia, New York City, and, eventually, the West Coast, solemn ceremonies and tributes eulogized the fallen colonel, and many thousands came to pay their respects. Herman Melville published a poem in his honor, as did young Willie Lincoln. Baker had, indeed, become a national martyr. To accuse such a man of having made mistakes that led to his own death and the deaths of so many brave soldiers just seemed wrong. A movement to point the finger of blame elsewhere soon began, and what better target than the general in charge on that fateful afternoon?[7]

Stone, meanwhile, was quick to pick up on the controversy that had begun to swirl around him. At the end of October, an unauthorized copy of his report to McClellan appeared in the press. On October 31, the *New York Times* noted that popular sentiment was alternating between praise for Baker and condemnation of Stone.[8] Two days later, he wrote to the assistant adjutant general to protest "the persistent attacks made upon me by the friends (so called), of the lamented late Colonel Baker, through the newspaper press." He felt it his duty to call these attacks to the attention of the army high command. Army policy forbade him from responding publicly, and, besides, he had too many duties reorganizing his command and tending to the wounded to issue denials to "every false statement . . . pronounced to be true." He intimated that Baker's supporters may have altered the bloodstained paper taken from the colonel's hat to make it appear as though he had been ordered to advance on Leesburg. Stone was particularly incensed at the rumor then current that upon receipt of this order, Baker exclaimed, "I will obey General Stone's order, but it is my death warrant."

"Shame upon them," he wrote, "to put false words in the mouth of the brave dead!" In reality, Baker had asked for the written direction and had seemed pleased to receive it. Stone then described how Baker precipitated the battle and then mishandled it, concluding that, "had his eye for advantage of ground in posting troops equaled his daring courage, he would have been to-day an honored, victorious general of the Republic, instead of a lamented statesman lost too soon to the country."[9]

The matter continued to fester. The discovery along the banks of the Potomac of the bodies of soldiers drowned in the retreat from Ball's Bluff added to the public's disgust with the battle and desire to punish whomever was responsible. Finally, in early

December, either at the request of McClellan, who was coming under increased scrutiny himself for Ball's Bluff, or on his own initiative to clear up growing confusion over what had transpired on October 21, Stone prepared a summary of the battle. He minced no words in a statement that he hoped would make it perfectly clear to all concerned just who had done what, and when:

> Stated concisely, the narrative would be this: General Stone directed Colonel Baker to go to the right and in his discretion to recall the troops then over the river or cross more force. Colonel Baker made up his mind and declared it before he reached the crossing place, to cross with his whole force. General Stone directed five companies to be thrown into a strong mill on the right of Ball's Bluff. Colonel Baker allowed these companies to be diverted to the front. General Stone sent cavalry scouts to be thrown out in advance of the infantry on the right. Colonel Baker allowed this cavalry to return without scouting and did not replace it, although he had plenty at his disposition. Colonel Baker assumed command on the right about 10 A.M., but never sent an order or messenger to the advanced infantry until it was pressed back to the bluff about 2:15 P.M. Colonel Baker spent more than an hour in personally superintending the lifting of a boat from the canal to the river, when a junior officer or sergeant would have done as well, the mean time neglecting to visit or give orders to the advanced force in the face of the enemy. No order of passage was arranged for the boats; no guards were established at the landing; no boats' crews detailed. Lastly, the troops were so arranged on the field as to expose them all to fire. While but few could fire on the enemy. His troops occupied all the cleared ground in the neighborhood, while the enemy had the woods and the commanding wooded height, which last he might easily have occupied before the enemy came up.[10]

Stone may have believed this explanation put to rest the controversy over who was to blame for Ball's Bluff, but he was badly mistaken. To his enemies, such statements merely added more fuel to their fire. Baker could not defend himself. He had died bravely leading his troops in battle, while his commanding officer sat safely on the Maryland shore. And now that same officer sought to escape criticism by placing responsibility for the debacle on their favorite colonel. It was up to loyal Union men to take up the cause in his behalf. If the army could not, or would not, discipline Stone for his incompetence, there were other ways of setting things right.

In addition to attacking his ability as a soldier, Stone's critics began hinting of disloyalty as well. They directed much of their hostility against the general's alleged admiration of Southerners and soft stance on slavery. It was true that Stone was by no means an abolitionist, and, like many other senior officers at the time, he believed the war should be fought to preserve the Union and not to end slavery. Emancipation as a war aim was not yet in the offing. His reputation as pro-Southern derived largely from statements he was said to have made praising certain Confederate officers. One such statement, reported by the pro-abolition *Chicago Tribune*, allegedly referred to the rebels as "gentlemen [who] are to be believed and relied upon . . . many of them are my intimate friends."[11] Charges also circulated that he ordered his troops to return fugitive slaves who had sought refuge in the Union camps. In one case he did have soldiers of the 20th Massachusetts escort two slaves who supposedly wished to return to their families (and masters). In this, as well as in other matters pertaining to slavery, however, Stone carried out not his policy but that of the general government and War Department.

Such legalisms, however, meant little to the rabid abolitionists. The *Chicago Tribune* charged that "nigger catching" had become Stone's "chief distinction." Some of the anti-slavery officers of the 20th Massachusetts also complained to their governor about Stone's conduct. On December 9, Governor John Andrew's military secretary dispatched a letter to Lieutenant Colonel Palfrey of the 20th Massachusetts, reprimanding the officer in charge of the slave-returning detail for compelling his "citizen-soldiers" to be "kidnappers of their fellow men." The governor further instructed Palfrey to prevent the men of his regiment from undertaking "any such dirty and despotic work in the future."[12] The letter was forwarded to Stone, who found it "extraordinary." He in turn sent it on to army headquarters so that McClellan could "devise measures which shall in the future prevent such unwarrantable and dangerous interference" in military matters. He went on to point out that the officer in question had not violated any regulations or state laws, and that governors could not issue "orders affecting the discipline of any regiment." From Stone's viewpoint, he did not have

Massachusetts, Michigan, Minnesota, New York, or Pennsylvania troops in his command—only United States soldiers.[13]

McClellan rightly responded to Andrew, echoing his general's sentiments. This touched off a bitter exchange between the governor and Stone regarding civilian meddling in his command. So rancorous did the dispute become that when the abolitionist Massachusetts senator Charles Sumner later degraded Stone on the floor of the Senate for his fugitive slave policy and alleged that by using his state's regiment to carry out those orders, he discouraged enlistments from that state, Stone exploded: "There can hardly be better proof that a soldier in the field is faithfully performing his duty than the fact that while he is receiving the shot of the public enemy in front he is at the same time receiving viterpuration [*sic*] of a well-known coward from a safe distance in the rear."[14] Given the customs of the time, his words could have been just grounds for a duel. Fortunately, cooler heads prevailed. Such an event could hardly have helped Stone since public opinion would have been overwhelmingly on the side of the senator.

New Yorker Roscoe Conkling also lashed out in the House of Representatives at what he considered to be "the most atrocious military murder ever committed in our history as a people," likening the battle to the Charge of the Light Brigade during the Crimean War. After accusing Stone of being "a martinet and not a soldier," he asked his colleagues, "Shall we proclaim indulgence for ignorance and incompetency, immunity for barbarous negligence, silence for military crimes?"[15] On December 10, at the behest of Conkling and other pro-war legislators, Congress passed a resolution creating an investigative body called the Joint Committee on the Conduct of the War. The joint committee sought to uncover information on the causes and consequences of recent federal defeats and the competence and loyalty of army leadership. Not quite a star chamber, its proceedings, nonetheless, were to be kept strictly confidential. Its membership consisted of pro-war senators and congressmen and was dominated by Republicans. At its head was the powerful abolitionist senator from Ohio, Ben Wade. Although Wade maintained that the joint committee existed only to gather information, render an opinion, and then let others in positions of civil and military authority decide what to do, it was clear that its real purpose was to extend congressional control over the rapidly growing army through intimidation and micromanagement.[16]

About the same time as it established the joint committee, Congress called upon the secretary of war, "if not incompatible with the public interest, to report . . . whether any, and if any, what, measures have been taken to ascertain who is responsible for the disastrous movement of our troops at Ball's Bluff."[17] When Secretary Cameron decided that furnishing such information would, indeed, not be in the public interest and declined the congressional request, an indignant joint committee elected to pursue its own probe into Ball's Bluff and General Stone. Besides inquiring into the battle itself, the joint committee addressed the allegations that Stone allowed Confederate spies through his lines, maintained better-than-average relations with rebel slave owners in the region, and had forced his soldiers to return runaway slaves to their masters.

The joint committee did not begin its work with Ball's Bluff; it initially examined what went wrong at Bull Run. In particular, it called into account the action, or lack thereof, of General Robert Patterson and his failure to contain the rebel forces under Joe Johnston. Testimony began on Christmas Eve and continued into March 1862. But by this time Patterson had been out of the army for several months, and, although the joint committee placed the blame for the defeat on him, few members of the press and fewer politicians were calling for his head. Likewise, the hapless General McDowell had long since been removed from command and replaced by McClellan. Attacking these has-beens would gain the joint committee little. McClellan and Stone, however, were still active and could be called to account for the disaster on the Potomac. Through them the joint committee might get in its licks against one of its main targets, the army's professional, West Point–trained officer corps. Many of the radical Republicans in Congress did not like the idea of a strong standing army, and particularly distrusted the United States Military Academy. Graduates of this institution tended to be elitists and Democrats who appeared to have more in common with their planter-class Southern foes than with loyal Unionists. After all, were not most of the Confederacy's leading generals West Point alumni? The joint committee and its supporters also believed that, despite their military bearing, men such as Stone and McClellan could not be counted on to wage the kind of aggressive war needed to bring the slavocracy to its knees.

Unfortunately for the joint committee, McClellan was still a bit too powerful and well liked by the army and the public to be attacked directly. On the other hand, Stone, while respected by many of his colleagues, did not have a large loyal following in or out of the service. And he had just suffered an embarrassing defeat. If the joint committee could not get at McClellan himself, it might be able to undermine him through one of his chief subordinates.

Chairman Wade gaveled the joint committee's investigation of Ball's Bluff to order on December 27. After Lander and McCall had given their testimonies, Stone was called on Sunday, January 5. He had requested to be examined then because the new Confederate general across the Potomac, D. H. Hill, was said to be deeply religious and not liable to launch an attack on a Sunday. Stone first briefly gave information about Patterson's role in the Bull Run campaign. Later, he returned to answer questions about Ball's Bluff and his policy on runaway slaves. Lander and McCall had given general, but not particularly damning, criticisms of Stone's conduct at the battle. The joint committee, however, was out for blood, and its tenor changed when Stone appeared. The members' interrogation became cynical and condescending. They asked hypothetical, "what-if" questions that Stone, who was there to provide facts, found difficult to answer. Although his anger simmered, Stone remained calm, placing the blame for the defeat squarely on Baker for his mishandling of the boats, his impulsive decision to send troops onto the bluff without having first viewed the scene of the battle, the poor deployment and management of his troops once there, and other errors. As he had been instructed by his superiors, Stone also deflected questions about McClellan's role and plans.[18]

Since most of his inquisitors were colleagues and friends of the late Senator Baker, Stone's testimony did him little good. When the questioning turned to the issue of Stone's stand on slavery, he stated that in returning runaways to their owners, he was simply following directives from the secretary of war and the laws of the state of Maryland, to which he added:

> The slaves that run away from the enemy and come over are got to my head-quarters as rapidly as possible; they are questioned

carefully, and all the information I can get out of them is taken. They are made as comfortable as they can be, and put to work in the quartermaster's department. . . . If they can take care of themselves, they have been allowed to do so the best way they could. If they have needed assistance, they have been fed and clothed and put to work by the quartermaster or commissary. I am not aware of any slaves coming over from the enemy lines having been given up to any claimant.[19]

As he was questioned further, Stone maintained that since martial law had not been proclaimed in Maryland, he had to allow civil authorities with proper warrants to come into his camps and remove persons, including runaway slaves. Only the property—and Negroes were considered property—of persons "in arms against the United States" would be retained.[20] This explanation was correct but it did not please joint committee members who held that the army should shelter slaves from Maryland, despite the fact that the state had not seceded and its citizens were not considered to be in rebellion.

Although not questioned about his dispute with Andrew and Sumner, Stone let his temper get the better of him by referencing the matter: "I might here say that vast injury is being done, insubordination is sown in the army, right and left, by the course pursued by newspapers and by public men in that respect." When asked what he meant by public men, he snapped back:

> The governor of a State, for instance, writes to the lieutenant colonel of a regiment reprimanding in the sharpest manner possible an officer of that regiment. And we cannot call in question the action of a senator or member of Congress on the floor of the Senate or House. But I have had in my own camps soldiers discussing in their tents the conduct of their general and the senator from their State, not knowing anything about the original circumstances, but simply discussing what their senator says of their commanding general. That is not a healthy state of discipline at all.[21]

Chairman Wade dismissed the statement out of hand, and the joint committee moved on to other matters before adjourning.

Although he had answered the questions to the best of his ability, Stone realized that the session had not gone well. A military court of inquiry would likely furnish him a better opportunity to protect his reputation. He asked for such a hearing, but headquarters denied the request. The last thing McClellan wanted was an airing of the Army of the Potomac's potentially dirty laundry. Stone was being sacrificed.

Following Stone's examination, the joint committee called a host of other witnesses, most of them subordinate officers serving under his command. Almost all of these men were volunteers who had been in the army for less than a year and disliked their commanding officer. They resented his stiff-necked discipline and strict adherence to military regulations and protocol. When quizzed, they took every opportunity to undermine Stone by attacking both his competence and loyalty. Much of their testimony was based on rumor and hearsay—tales they had heard from other soldiers or read in the newspapers of Stone's Southern sympathies, popularity among secessionists, trafficking with the enemy, ordering Baker to march on Leesburg, or returning terrified runaway slaves back to their rebel masters. Some stated they would refuse to serve under the general and that such sentiments were common among the ranks.[22]

Occasionally, witnesses defended Stone, as in the testimony of Captain Clinton Berry. When questioned about the opinions of the officers and men regarding their general, he responded, "I think that the better informed portion of the army there, that is, the educated men, have great confidence in General Stone." Asked if these were "West Point educated men," he stated that they were, but that:

> In regard to my own opinion of General Stone, I think he is certainly one of the most accomplished soldiers and gentlemen that I ever had the pleasure to meet. His management of his division shows that. He is constantly bringing into effect changes to make our volunteers equal to regulars. He follows the army regulations a little to [sic] strict for volunteers, perhaps, but I think it is out of regard to our own interests.[23]

When such remarks were made, the chairman more often than not either ignored them or switched the questioning to subjects that might elicit negative comments about Stone.

On January 27, while the investigation was still in progress, members of the joint committee called upon the new secretary of war, Edwin M. Stanton, to present him with the "evidence" against Stone's loyalty that it had collected. The secretary, a strong opponent of secession and slavery, accepted the findings. He then went beyond the joint committee's recommendation that the army itself investigate the general, and the next day issued an order to McClellan for Stone's arrest and close confinement.[24]

As in other instances, McClellan was slow to act on Stanton's order. He knew Stone had not been primarily responsible for Ball's Bluff and that his fidelity to the Union was solid. At the same time, he believed the joint committee was searching for a victim and his beleaguered general had not mounted the strongest defense. If Stone were to fall, perhaps he would be next. If, however, Stone had one more chance to explain himself to Wade and the others, the arrest might be cancelled. McClellan persuaded Stanton to delay the order and allow Stone to testify again, and went to the joint committee with the same request. It agreed, not out of deference to McClellan or sympathy for Stone, but perhaps anticipating an opportunity to degrade both generals even further.[25]

McClellan now had to arrange for Stone to return and defend himself before the joint committee. He sent a message for him to report immediately to headquarters. Stone, who had prepared and sent the commanding general ideas for another attack on Leesburg, first thought he was going to discuss those plans. The order arrived at nine P.M. Stone jumped up from his camp desk and within fifteen minutes was on the road to Washington. Upon arrival, however, McClellan informed him that he was relieved of his command (at least temporarily), and that the joint committee was now questioning his loyalty. He was being called back to testify again. Stone was dumbfounded. To be accused of incompetence by a group of civilian politicians with little or no knowledge of military affairs was one thing. To have his loyalty called into question was quite another matter.[26]

On Friday, January 31, Stone entered the lion's den once more. Instead of being able to review the allegations against him, however, he had to listen to Chairman Wade summarize the charges and state that "in the course of our investigations there has come out evidence . . . which may be said to impeach you." Stone once more denied responsibility for the defeat at Ball's Bluff, citing Baker's mistakes. The chair then accused him of improper communication with the enemy, to which he responded angrily: "That is one humiliation I hoped I should never be subjected to." In a long and detailed statement, Stone refuted the "utterly and totally false" charge, reminding Wade and the others that as commander of the militia in the District of Columbia during the secession crisis, he held the safety of everyone in Washington, including the president and Congress, in his hands. If he were a traitor, what better time to have made his treason known? In an emotional outburst, he contended, "And now I will swear that this government has not a more faithful soldier; of poor capacity it is true; but a more faithful soldier this government has not had from the day General Scott called me . . . up to this minute."[27]

Later in his testimony, Stone recounted his service to the Union and again dramatically professed his loyalty:

I have, so help me Heaven, but one object in all this, and that is to see the United States successful. I have from the first day of January of last year till this day hardly been out of my clothes. After the 7th of June, the moment this capital was safe, I was sent away, I was kept here until it was safe, and my thirty-three companies of men made it safe. I say it without contradiction, that the thirty-three companies of men under my command held this capital safe. I claim it for them, not for me. It was no sooner safe, no sooner occupied, than I was sent up the canal to guard the outposts of Washington with some of those very men, and other men from the United States, and from the day I then left Washington until this day, I have been upon the outposts, with the exception of three short visits I have made here by order, and except during those visits here, I have gone to bed, that is got into my blankets, every night without undressing; from the 7th of June to this day, while in the field, the enemy never could have surprised me at any time without my being dressed and outside of my tent in one half minute. The most I have ever done has been to pull off some large boots which I could put on again in an instant. If you want more faithful soldiers you must find them elsewhere. I have been as faithful as I can be.[28]

Wade, the joint committee, and Stanton were not impressed by this curious but impassioned statement.

Stone's second round of testimony ended with inconsequential discussions of why he had not used artillery against rebel canon and fortifications near Leesburg—from the balloon he could see that the earthwork containing the enemy's guns were out of range and the other remained unoccupied; why he had not destroyed certain flour mills in Virginia—such acts would have accomplished nothing militarily and only provoked retaliation; and why he had allowed farmers who had plantings on islands in the Potomac to gather their crops—the harvest was to supply the Union army. Chairman Wade did not bother to acknowledge the explanation. Stone believed he had done all he could to defend his reputation as a soldier and a loyal American. He now looked forward to being restored to command and returning to the Corps of Observation. D. H. Hill still had a force nearby in Virginia, and he was anxious to take the offensive against it. But McClellan kept Stone in the capital.

Just why would soon become clear.[29]

NADIR

B Y all accounts, it was a "brilliant spectacle." On the evening of February 5, 1862, Mary Todd Lincoln held her first invitation-only reception in the White House. The president was there as he and the First Lady greeted their guests in the East Room. The opulent Red, Green, and Blue parlors, festooned with exotic flowers, also were open for diplomats, cabinet members, judges, generals, and politicians, along with their wives, to mingle and make small talk. McClellan appeared resplendent in his full dress uniform. The Marine Band played until eleven o'clock, when a supper, accompanied by the finest wines and champagnes, was served by Malliard, one of New York City's leading caterers. The table featured a confectionary model of the steamer *Union*, and on a side table sat a representation in sugar of Fort Sumter. The festivities lasted until three in the morning, when the band played "The Girl I Left Behind Me" as guests departed. Stone and Maria Louisa attended and enjoyed the event, little knowing that seventy-two hours later he would be under arrest.[1]

Three days after the White House reception, Stone was engaged in a late-evening meeting with some colleagues at Willard's Hotel. Over brandy and cigars they reviewed the latest detailed map of the Confederacy. The gathering broke up about midnight, and Stone walked home. Approaching his residence, he noticed an army officer waiting out front. He recognized the man as George Sykes, an old friend and regular army major in Stone's 14th Infantry Regiment. Sykes had recently been promoted to brigadier general in command of Washington's city guard, and Stone greeted him cordially and congratulated him on his recent advancement. But he could see something was not quite right. The clearly uncomfortable Sykes stated that he had a "disagreeable duty" to perform. He was there to arrest him. An astounded Stone questioned why he was being arrested. Sykes professed not to know, but suggested he go inside and change into civilian clothes, as he was being sent to Fort Lafayette. Once again Stone was flabbergasted, "Fort Lafayette is where they send secessionists!" All Sykes could reply was, "Yes."[2]

Stone entered the house, changed into a suit, and reassured Maria Louisa that everything was going to be all right. He would contact McClellan who would rectify the mistake. When he emerged through the front door, he noticed a squad of riflemen that had been concealed in the shadows. With Sykes at Stone's arm, and flanked by the soldiers, the party marched to the Chain Building, headquarters for the city guard. There Stone surrendered his personal possessions and was placed in a lieutenant's bedroom for the night. An armed sentinel stood outside. Oddly enough, "although for the first time of my life in any arrest of any kind," he "slept soundly until daylight."[3] The following morning, Stone requested and received writing materials. He dispatched a message to McClellan's adjutant informing him of his arrest and asking to view and respond to the charges that had resulted in his being detained. There must have been some "strange misunderstanding" that the army high command could clear up quickly so that he might return to his division.[4]

McClellan, of course, already knew of Stone's arrest. He had issued the order. In fact, he had known for some time that the unfortunate general was to be imprisoned. He had received Stanton's order for the arrest, but had been reluctant to serve it. The evidence against Stone was flimsy, but disobeying the secretary of war in this matter could be harmful to his own career. McClellan found a way out of this dilemma, however, when detective Allan Pinkerton brought him a refugee from Leesburg who spun a hearsay tale involving Stone's alleged improper communications with the enemy and Confederate officers' praise of Stone's character and gentlemanly qualities. McClellan interviewed the refugee personally and then took this new "evidence" to Stanton. The secretary reaffirmed the arrest and instructed McClellan to get on with it. Convinced that he had done everything he could for Stone, the commanding general wrote out the order to the provost marshal for Stone to be taken into custody, held incommunicado, and transported to Fort Lafayette. At first he added "to await trial," but later crossed it out.[5]

Stone spent the morning after his arrest still in the lieutenant's room. In the afternoon, a "gentlemanly lieutenant" arrived and took his parole that he would not try to escape. Stone appealed to Sykes that all McClellan had to do was order him to report to Fort Lafayette and "his order would be effective as any guard." The sympathetic Sykes responded that if it were up to him, he would allow Stone to go there unattended, but "somebody" in authority had insisted he be taken under guard as "a dangerous person."[6] Stone was given back the contents of his pockets, and the "gentlemanly lieutenant" and two detectives then took him to the Baltimore and Ohio Railroad depot to begin the journey to Fort Lafayette. An embarrassing incident followed when the train reached Philadelphia. Their passage required the purchase of new tickets for the remainder of the trip. Neither the lieutenant nor the detectives had enough money to cover his fare, so Stone ended up paying his own way to prison.[7]

Early the next morning, the party appeared at the gate of Fort Hamilton on the Brooklyn side of New York Harbor. There Stone was turned over to Lieutenant Colonel Martin Burke, an old "friend and comrade" from their Mexican War days, who commanded Fort Lafayette as well. Burke had the order for his incarceration, but could not believe it. "In the name of God, Stone, what does this mean?" he exclaimed. Stone responded quietly that he did not know. The colonel stated that he would have to carry out McClellan's instructions, but he was not sure he could provide his prisoner with the commanding general's request for "the comfort due his rank," at Fort Lafayette.[8]

The United States Military Academy in West Point around the time of Charles P. Stone's attendance. The plain where plebes learned the basics of military life is in the foreground. (*Library of Congress*)

General Winfield Scott overseeing the bombardment of Vera Cruz, Mexico, in March 1847. Deadly mortar rounds arc into the city and the fortress of San Juan Ulua in the background. Despite this artist's rendition, the infantry played a relatively minor role in the siege. (*Library of Congress*)

The Battle of Molino del Rey, September 8, 1847. The wrecked cannon shows the danger faced by artillery crews such as the one commanded by Stone. As in many popular illustrations of the War with Mexico, the American soldiers are depicted in full dress rather than in their blue and gray field uniforms with low billed caps. (*Library of Congress*)

The US military post at Benicia, California, as it appeared when Stone and his ordnance company arrived. (*Author*)

The view of Suisun Bay and the Ordnance Depot at Benicia some time after Stone's departure. The first sandstone store house built by Stone can be seen in the foreground. (*Library of Congress*)

The San Francisco building that housed Lucas, Turner & Company where William Tecumseh Sherman and Stone conducted business in the 1850s. Following the Civil War, Sherman recommended Stone to the Khedive of Egypt to modernize his army. (*Author*)

Charles P. Stone in civilian dress in the 1850s. (*Library of Congress*)

Ignacio Pesqueira, Stone's implacable Sonoran adversary, in later years.

Port of Guaymas, gateway to Sonora, Mexico, about the time Stone and his surveyors sailed in aboard the *Manuel Payno* in April 1858. (*National Archives*)

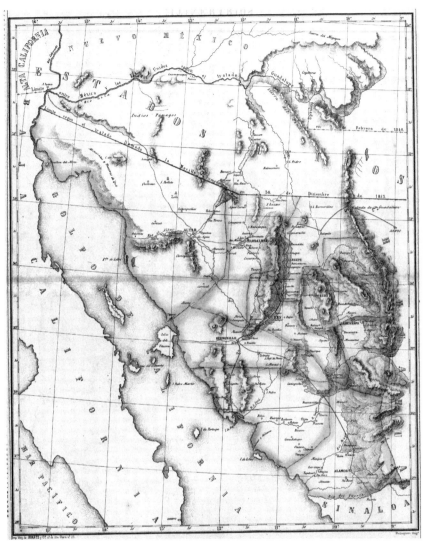

A map of Sonora, Mexico, published in 1858, and showing the road likely taken by Stone's surveyors from Guaymas, through Hermosillo and Altar and then north to the border. This also would have been the route of Captain Richard S. Ewell on his mission to secure the readmission of Stone and his men to Sonora after they had been forced to flee. Sonora's governor Pesqueira had expelled Stone's mission claiming that the survey party was a prelude to possible American purchase or annexation of the state. (*David Rumsey Historical Map Collection*)

A view from a balloon of Washington City in May 1861. The Capitol dome is under construction in the foreground, while the unfinished Washington monument and the Long Bridge to Virginia appear in the distance. Earlier in the year, Stone had only a handful of troops with which to defend the city against possible secessionist insurrection or Confederate invasion. (*Library of Congress*)

Left to right: Charles P. Stone in uniform as a brigadier general of volunteers. (*National Archives*) General Winfield Scott as he appeared at the outbreak of the Civil War. He commanded Stone in Mexico and during the "siege" of Washington in 1861. Senator Edward D. Baker, a close associate of President Abraham Lincoln. Baker's impetuosity and lack of military skill at the battle of Ball's Bluff cost him his life, the Union army a victory, and Stone his career and freedom. (*Library of Congress*)

Union troops desperately retreating from Ball's Bluff, Virginia, to the Potomac River on October 21, 1861. (*Library of Congress*)

A dramatic rendering of the slain Colonel Edward Baker being carried from the field of battle at Ball's Bluff. (*Library of Congress*)

Left, Thaddeus Lowe's balloon *Intrepid*; right, an observation balloon being filled with gas. Stone employed the *Intrepid* to observe the Confederates across the Potomac River after the battle of Ball's Bluff. The general made occasional ascensions to view enemy fortifications and troop movements for himself. (*Library of Congress*)

In the weeks after Ball's Bluff popular sentiment fueled by inaccurate information alternated between praise for Baker and condemnation of Stone. The discovery along the banks of the Potomac of the bodies of soldiers drowned in the retreat, illustrated above, added to the public's disgust with the battle and desire to punish whomever was responsible. (*Library of Congress*)

Forbidding Fort Lafayette prison in New York harbor where Stone was incarcerated in February 1862, as viewed from Fort Hamilton. Stone was imprisoned without official charges or a trial. (*Library of Congress*)

Left, Ohio Senator Benjamin Franklin "Bluff" Wade, chair of the Joint Committee on the Conduct of the War and Stone's nemesis. Center, California senator James A. McDougall authored a clause in a military appropriations bill that required an officer must be furnished with a copy of the charges against him and be brought to trail within thirty days, effectively freeing Stone. Right Major General Nathaniel P. Banks requested the reinstated Stone to be chief of staff of the Department of the Gulf in May 1863. Within a year, however, the two men had become bitter enemies. (*Library of Congress*)

The battle at Pleasant Hill, Louisiana, April 9, 1864. Although the rebels appear to be getting the worst of it in this illustration, General Banks (lower right on horseback) and the Union army later "skedaddled" to safety at Grand Ecore, despite a gallant charge by Stone. The defeat at Pleasant Hill effectively ended the Union's Red River campaign. (*Library of Congress*)

Ismail "the Magnificent," Khedive of Egypt. He hired Stone in 1870 to be his army's chief of staff, giving him a second chance. In return, Stone served him loyally for nine years. (*Library of Congress*)

"Stone Pasha" in uniform as lieutenant general of the Egyptian Army. In addition to his Egyptian decorations, Stone wears a Grand Army of the Republic badge.

The Citadel in Cairo, headquarters for the Egyptian general staff during Stone's tenure. (*Library of Congress*)

Lieutenant Colonel Ahmed 'Urabi led an unsuccessful nationalist uprising in 1881–1882 against foreign influence over Egypt, paving the way for a British protectorate and ending Stone's service there.

President Ulysses S. Grant (center in sun helmet) and his wife Julia visiting the ruins at Karnak, Egypt, in 1878. Grant had known Stone since his West Point days. He would later be influential in the appointment of Stone to be chief engineer for constructing the base and pedestal of the Statue of Liberty. (*Library of Congress*)

Stone's loyalty to Ismail extended to serving his son and successor, Khedive Tewfik. After leaving Egypt, Stone began work on the Statue of Liberty.

Stone arranged a cornerstone laying ceremony for the Statue of Liberty on August 5, 1884. A steady rain made umbrellas a necessity. (*National Park Service*)

Harper's Weekly illustrated the pedestal under construction. The inset is Frédéric Bartholdi, the sculptor-designer of the Statue of Liberty. Outwardly, Bartholdi expressed confidence in Stone, but privately he was disappointed in the slow pace of work on the pedestal. (*Library of Congress*)

By the end of 1886, Stone and his workmen had completed the pedestal on which the statue would stand. (*Library of Congress*)

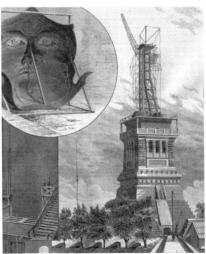

The cover of Scientific American featured the statue, with framework designed by Gustave Eiffel, under construction. Stone had devised the method of anchoring the statue's frame to the pedestal. (*Library of Congress*)

Photograph taken of the Statue of Liberty's inauguration ceremony on October 28, 1886. A heavy mist combined with smoke and steam from a flotilla of ships and boats obscures the statue from onlookers lining Manhattan's waterfront. (*Library of Congress*)

The enduring symbol of American liberty. A museum in the base contains a plaque and exhibit acknowledging the under-appreciated contribution of Charles Pomeroy Stone to the nation. (*Elcobbola/Wikipedia Commons*)

Comfort was hardly a word associated with Fort Lafayette. Located offshore in the narrows between Staten Island and Long Island, it had been built between 1812 and 1819 for the defense of New York Harbor. With sandstone walls twenty-five to thirty feet in height and eight feet thick, Fort Lafayette appeared imposing and grim to those facing incarceration there. Some observers referred to it as the "American Bastille," although a writer for the *New York Times*, with tongue in cheek, maintained that it was "more like a hotel than anything else, where the proprietor is rather strict." Both prisoners of war and civilians accused of conspiring with the Confederacy had been housed there since July 1861. On the morning of February 10, the fort prepared to receive its most famous inmate.[9]

Stone was rowed the quarter mile from Fort Hamilton to the gloomy island of Fort Lafayette. He again surrendered what he had left in his pockets and enjoyed a good breakfast from the officers' mess onshore prior to being assigned quarters. Most prisoners were crammed into casemates that had been converted into barracks. In Stone's case, however, he was given a private cell in a casemate formerly occupied by enlisted guards. His principal jailor, Lieutenant Charles Wood, described the general's arrival and accommodations:

> He was immediately placed in one of the most comfortable rooms of the quarters and supplied with everything possible to give him in the way of furniture, such as iron bedstead, mattresses, pillows, linen, blankets, towels, tables, chairs, washstand, washbowl and ewer, looking glass, lamp, lamp oil, water buckets, &c., and a soldier detailed to wait on him. A sentinel was placed outside his door and upon the porch and he was not allowed out of his room except to go to the water closet . . . at which time he was accompanied by one of the guard. The troops of this command were specially charged that they must show him the respect due his rank and no instance of disrespect ever came to my knowledge.[10]

Wood had a reputation among some prisoners as being "brutal," but Stone found him to be polite and considerate. Unlike his fellow inmates, who dined primarily on beef, pork, potatoes, and rice twice a day, he did not take his meals from the inadequate prison kitchen but continued to eat food prepared by the officers' mess at

Fort Hamilton and rowed across to him. Stone may also have been able to use his background in ordnance to curry favor and obtain some "delicacies of the market," for the fort's ordnance sergeant was also in charge of the mess. He recalled later that during his incarceration he received "many kindnesses" from the soldiers' families as well.[11]

At first, Stone was held with no communication to or from the outside world. Within a few days, however, he began obtaining newspapers. In their pages he read of the "frightening crimes" of which he had been accused. Angered by what he considered outright lies, he sent messages to the adjutant general asking for copies of the charges against him, access to his official and personal papers, and for a hearing. No replies were forthcoming from Washington. McClellan, had, in fact, ordered Stone's papers collected and sealed from view by anyone, including their creator, and the last thing Stanton wanted was for Stone's case to appear in even a military court.[12]

All of Stone's incoming and outgoing correspondence was reviewed and censored by the officers at Fort Hamilton. Questionable letters were referred up the command chain for approval or return. Initially, McClellan approved all visitors, but he soon delegated this responsibility to Burke. Just one week after his arrival at Fort Lafayette, Stone's sister received permission to visit him. His uncle-in-law, Commander Albert Gallatin Clary, stationed at the nearby Brooklyn Navy Yard, received similar permission in late February, but others whom McClellan, and later Burke, deemed unacceptable for whatever reasons, were denied access to the general. Some with political connections appealed to Stanton, but usually without success. Joseph H. Bradley, a prominent and controversial Washington attorney and friend of Stone's, for example, tried to get a letter to the general and then to visit with him, but was turned down flat by the secretary in both attempts.[13]

One thing his jailors and few visitors all noticed was that Stone's health had begun to decline. Fort Lafayette was an uncomfortable and unhealthy place in which to be imprisoned. Casemate cells were dark, cold, and damp, and Stone had limited opportunities for exercise. His day began at dawn, when the cell doors were unlocked. He could then spend an hour in the open area of the fort. This was followed by breakfast and an opportunity to visit the casemate rooms of the other prisoners, although—because he still was trying to demonstrate his loyalty—Stone likely did not fraternize much with his fellow inmates, many of whom were well-known secessionists. After a day of reading, writing, and smoking in his quarters, he was given another hour outdoors between five and six in the evening. He then returned to his cell to await dinner. Lights-out came at nine o'clock. This confinement soon began to tell on Stone's health. When McClellan learned of his deterioration, he ordered Burke to allow him increased opportunities for outside exercise. Burke tried to comply, but the environment of Fort Lafayette continued to take its toll. Near the end of March, word of Stone's decline had reached Stanton, who directed the surgeon assigned to the prison to examine the general and submit a report on whether he needed a change in his exercise routine or a new location of imprisonment altogether. Within a couple of days, the surgeon responded that "the present limits to which Brigadier-General Stone is now confined are insufficient for the preservation of his health." He required more open-area activity than could be provided at Fort Lafayette. The prisoner had to be moved, and soon.[14]

The secretary of war now had to make a decision. He wanted Stone out of the way, not dead. A general who passed away while in a military prison might turn from object lesson to martyr in the eyes of the officer corps. On March 28, therefore, Stanton authorized Stone to be moved to Fort Hamilton, where he would have better "opportunity for taking the air and exercise." Although still a prisoner, Stone was granted a conditional parole that allowed him to leave Fort Hamilton from time to time and stroll about the adjacent village. On all such visits, a commissioned officer had to accompany him, and he had to be back by retreat. The secretary of war did, however, deny Stone a parole that would have allowed him to visit New York City, and his request to have Maria Louisa reside with him at the fort.[15]

During the seven weeks that Stone had languished at Fort Lafayette, events were occurring on his behalf outside the prison walls. Word of his arrest had sent shockwaves throughout the army's higher-ranking officers. They could not believe that a distinguished general's reputation could be so easily ruined. Winfield Scott's response to the news summed up the opinions of many of his fellow generals: "Colonel Stone a traitor! Why if he is a traitor, I am a traitor, and we are all traitors." Nevertheless, many of his colleagues, McClellan included, while they privately sympathized with Stone did not rush to his defense, lest their own careers be put in jeopardy. It was a dangerous time to be a general in the Union army.[16]

Civilian support was another matter. Stone's brother-in-law, Henry Melville Parker, a Boston lawyer and former secretary of that city's prestigious Athenaeum, became his attorney and chief advocate. Soon after Stone's arrest, he journeyed to Washington to confer with McClellan's adjutant and arrange a prompt military tribunal. He got nowhere. On February 19 and 20, he met with Stanton to discuss the case. The secretary matter-of-factly outlined what had transpired and told Parker there was nothing extraordinary about the matter. Parker responded with a letter to Stanton in which he maintained that Stone's situation was, indeed, extraordinary in every respect. He pointed out that prior to his arrest he had never been censured or disciplined by any military superior, including Stanton himself. Prior to his arrest, Stone had repeatedly asked for a court of inquiry regarding Ball's Bluff but had been turned down each time by the War Department. Stone's incarceration without a preliminary examination also was extraordinary, as was the fact that twelve days had passed and the accused, his family, and his counsel still had no information about the accusations against him. Parker maintained that when Stone is eventually found to be innocent, "the one worthy reparation which the Government can make . . . giving him the opportunity once more to prove his innocence . . . in the hot work of war" will likely not be possible because the rebellion will by then be over. Finally, he protested Stone's imprisonment, since the public regarded anyone sent to Fort Lafayette as guilty.[17]

Parker's meetings and letters did not win Stone a hearing or access to the charges against him or access to his papers, which remained under McClellan's lock and key, but it did gain the attorney access to his client. Together Stone and Parker made repeated appeals for a court of inquiry or even a court-martial to clear the general's name. After all, the circumstances under which Stone had been imprisoned ran counter to the army's Eightieth Article of War, which provided that at the time of incarceration the commandant of the prison must be presented with "an account in writing . . . of the crime with which the said prisoner is charged." Burke claimed he had never received such an account. Stone was also being held in clear violation of the spirit, if not the letter, of the Seventy-ninth Article of War, which stated that "no officer or soldier who shall be put in arrest shall continue in confinement more than eight days, or until such time as a court martial can be assembled." All of Stone's requests were met with either silence from the War Department or replies that "they [the charges against Stone] will be filed in good season."[18]

Stone and Parker were not done, however. They adopted another strategy. On April 13, the general applied for permission to serve in the Army of the Potomac's Peninsula Campaign. Stone was still a commissioned officer who had never been tried or convicted of any violation of regulations. His request served as a demonstration of his loyalty. Despite being improperly arrested and imprisoned, he still was anxious to serve his country. The request was given to Burke who approved transmittal to Washington. When he did not receive a response, Parker contacted the adjutant general's office, but to no avail. The War Department clearly wanted to keep the general under wraps. As day after dreary day passed with no signs of progress, a dispirited Stone finally had to admit, "I am in a complete muddle."[19]

By the spring of 1862, it had become clear that navigating through normal military channels would not free Stone. A political solution, however, held more promise. Stone and California's junior senator, James McDougall, had known one another in the Golden State during the 1850s. Although it was unlikely that the pious and gentlemanly officer and the boisterous, hard-drinking senator were close, McDougall took up the cause at the request of Parker and Stone's supporters in California and Massachusetts. Perhaps he saw it as a way to help a fellow Democrat, or possibly as a way to needle the radical Republicans in Congress. Whatever his motivation, he made three trips to the War Department to inquire about the general's incarceration. When he could not get an answer, he joined with his fellow senator from California, Milton Latham, and Representative A. A. Sargent in a letter to Stanton questioning why Stone had been imprisoned without trial, "a proceeding which seems to us most extraordinary." No reply was

forthcoming, so on April 12 McDougall introduced a Senate resolution calling for the secretary of war to provide a full accounting of
the facts surrounding Stone's arrest and imprisonment. Specifically,
the resolution asked who had authorized that he be taken into custody and whether at the time the general was subject to the
Articles of War. If so, was he arrested for violating them? It asked
whether any charges had been preferred against Stone, or if charges
and specifications had been drawn up so that the matter could be
referred to the Judge Advocate General or other magistrate.
McDougall's resolution went on to inquire whether Stone had
asked for a speedy trial since, as the war progressed, key witnesses
might be dead, held captive, or otherwise unable to testify. Once
again, what had been the War Department's response to this
request? Then came the key questions. Had Stone asked to see the
charges and evidence against him, and, if so, had they been communicated to him? If not, why not? And, was the case of General
Stone being handled in accordance with the army's own regulations?[20]

In speeches over the next two days, McDougall presented an
account of Stone's previous service to the republic, emphasizing
the general's patriotism and devotion to duty, along with the
unjustness of his arrest "by tyrant's law" and his continued imprisonment without trial. He laid bare the War Department's mistreatment of its general, but Stanton and his allies in the Senate did not
intend to go down without a fight.[21]

McDougall's resolution and orations touched off a particularly
acrimonious exchange between the California senator and Ben
Wade. The Californian reiterated points from his resolution and
previous addresses, and cited precedents from English military law
and American history pertaining to the rights of officers. Wade
countered with the assertion that during the English Civil War, the
traditional guarantees to officers were often disregarded. In
response to Stone's previously unblemished record, the sharp-
tongued Wade offered the example of Lucifer, "once a bright angel
in heaven, but he fell, and he has not been much honored in that
quarter since."[22]

The debate then turned to the Joint Committee on the Conduct
of the War. McDougall cited its abuses and likened it to the Holy
Office of the Inquisition. Wade tried his best to defend the com-

mittee as a body devoted to promoting competence, efficiency, and loyalty within the ranks of the army. Nevertheless, the resolution, amended to direct the questions to the president rather than the secretary of war, was adopted.[23]

Lincoln responded with a masterful example of political obfuscation, if not downright untruthfulness, that shed little light on Stone's case:

> I have the honor to state that he was arrested and imprisoned under my general authority, and upon evidence which, whether he is guilty or innocent, required as appears to me, such proceedings to be had against him for the public safety. I deem it incompatible with the public interest, as also, perhaps, unjust to General Stone, to make a more particular statement of the evidence. He has not been tried because, in the state of military operations at the time of his arrest and since, the officers to constitute a court-martial and for witnesses could not be withdrawn from duty without serious injury to the service. He will be allowed a trial without any unnecessary delay; the charges will be furnished him in due season, and every facility for his defense will be afforded him by the War Department.[24]

Despite the president's assurance, Stone was denied a trial and the evidence against him still withheld. As spring passed into summer, McDougall continued his effort to gain the general's release, but the case attracted little new attention. Stone may have become somewhat of a *cause celebre* among the army's senior officers, but the monumental events of the Civil War distracted the general public's interest from the plight of one disgraced soldier.

In midsummer 1862, however, Stone's case took a decidedly favorable turn. McDougall found a loophole in a military appropriations bill, and inserted a clause providing that an officer placed in custody must be furnished with a copy of the charges against him and brought to trial within thirty days, or he must be freed. This wording applied directly to Stone. The bill passed and was signed into law on July 17. Although he had no intention of bringing the general to trial, a vengeful Stanton kept him locked up for the full thirty days. Finally, on August 16, a smiling Colonel Burke presented Stone with a telegram from the secretary of war ordering his release, on grounds that "a court could not be convened within the time prescribed by law."[25]

After spending 189 days in confinement, Stone was free, but the matter had not yet closed. He immediately telegraphed Washington his receipt of the release and asked for orders. He was, after all, still a brigadier general of volunteers and naively believed a position would be waiting for him. The Union needed generals, particularly brigadiers. To resign would be to admit defeat and his name and reputation would be sullied forever. After remaining at Fort Hamilton for twenty four hours and hearing nothing back, he left for New York City. There he stayed another two days anticipating an assignment. When none came, he made his way to the Capital to be reunited with his family, report for duty, and clear his name. He again applied to McClellan to serve with the Army of the Potomac, "if only as a spectator," but did not receive a reply. He contacted the adjutant general's office, but was told that his name had not been referred for active service by the secretary of war. In 1863, the new commander of the Army of the Potomac, Major General Joseph Hooker, actually asked that Stone be assigned as his chief of staff, but this, too, was denied by the War Department with the standard line that such requests were "not . . . in the interests of the service." In only one instance was Stone briefly given orders. A clerk likely saw his name on a list of available generals and directed him to preside over a court-martial. After just one day as president, however, the War Department curtly relieved him from duty.[26]

Stone's attempts to get to the bottom of why he was arrested met with similar frustration. Deciding to start at the top, soon after arriving in Washington he asked for and received an interview with Lincoln. When asked by the general about the particulars of his incarceration, the president reiterated that the arrest was done under his "general authority," but not by him directly. He then stated that never had he believed Stone to be a traitor, and referred him to new Union general-in-chief Henry Halleck for further assistance. Halleck was of little help. He belabored the obvious by telling Stone that he was no longer under arrest and there were no charges pending against him. Other than that, he really did not know much about the case and had no more useful information. The adjutant general, professing ignorance, continued to put him off as well. Stone finally could not take being "bandied about from one authority to another," any longer without learning a single fact about his incarceration and abandoned the quest.[27]

At this point, the joint committee stepped in and granted him another appearance. On January 27, 1863, Stone had at last the opportunity to view the testimony against him and offered a lengthy and able defense. When asked why he had not given similar answers before, he replied, "If the chairman will remember, the committee did not state to me the particular cases . . . I gave general answers to general allegations." Stone also maintained that McClellan's staff had instructed him not to be specific when questioned. Freed from this restriction, he fully accounted for his actions before, during, and after the defeat that had landed him in prison. As he left the hearing room, the scapegoat of Ball's Bluff finally felt vindicated.[28]

The joint committee, however, did not see the proceedings in quite the same way. On April 6, 1863, it published its report on Ball's Bluff. It included all of the testimony as well as a summary, which left much of the responsibility for the debacle up in the air— a case of *he said, he said*. To clear Stone's reputation was just too much for the radical Republicans. Readers would have to make up their own minds about whether Stone was to blame, or Baker. The report further claimed rather disingenuously that the committee knew nothing about the general's arrest, except what its members read in the newspapers. As for the War Department's failure to furnish Stone with the charges against him, "why his request was not granted your committee have never been informed."[29]

During this time of professional frustrations, Stone's personal life took a tragic turn as well. In October 1862, his mother-in-law, Esther Phillipson, died, and on February 27 of the following year— just a month after his final appearance before the joint committee—Maria Louisa also passed away. She was thirty-one years old. As a new widower with a child to support, the general had to work through his grief and obtain some kind of an active-duty assignment. He haunted the halls of the War Department waiting for something to turn up. At last, in early May, Nathaniel Banks, now commander of the Department of the Gulf, applied for his services, and, for a change, Stanton and Halleck reluctantly approved. Neither man had much regard for Stone as a warrior, but he did have a reputation as an able administrator, and that was just what Banks needed. Besides, by this time the ever-present Stone was becoming somewhat of an embarrassment, and what better way to get rid of him than to ship him a thousand miles away to New Orleans. The Department of the Gulf was somewhat of a backwater and had a reputation as a dumping ground for disgraced officers. For his part, Banks claimed to have confidence in Stone's zeal and ability and made himself responsible "for his conduct in the future."[30]

Banks had been in New Orleans for almost six months, taking over from General Benjamin Butler, whose brusque manner in dealings with foreign consuls and his dictatorial management style made him a problem for the Lincoln administration. The politician Banks, on the other hand, brought an air of civility and decorum to the Department of the Gulf and sought conciliation with civilians in the region, particularly the planter class. This gained him some popularity, but the expectations that he would administer both civil and military matters while at the same time launching a campaign to assist with the capture of Vicksburg proved to be a challenge. Treasury agent George Denison first welcomed Banks as a change from the corruption and inefficiency of the previous regime, but soon determined the new commander to be "no judge of men." With a staff that was mostly inexperienced and/or inept, Banks likely realized he required some help when he asked for Stone's services.[31]

The generals had known one another since the summer of 1861, but it was hard to imagine two more different men. Stone was the consummate professional soldier, disciplined, trained, and experienced. Banks was a self-made politician who had risen from "bobbin-boy" in a textile mill to Speaker of the House of Representatives. On the other hand, he could not even understand a military map and did not know the first thing about leading an army in battle. In addition, he had a difficult time accepting advice, particularly from those he considered West Point martinets. Then there was Ball's Bluff. Banks believed he had rescued Stone, though Stone believed he had saved himself while Banks stood around doing practically nothing. Nevertheless, Banks needed a military administrator, and Stone could not afford to be choosy when it came to an assignment.

By June 1, Stone had reported for duty in New Orleans, a city that even his well-traveled eyes must have found exotic. Although part of the United States for six decades, its creole culture of the

Spanish and French still exercised a major influence in society, language, and architecture. It was likely the most foreign place in the nation, and one in which class distinctions remained strong, but racial and ethnic lines blurred, "a light, mobile, effervescent" place where "the French and demi-French, the Creole, the Octoroon, the Quarteroon, the Mulatto and the sable skinned meet and jostle one another" on a daily basis.[32]

Stone did not have much time to take in the sights, sounds, tastes, and society of the Crescent City. He hurried to join his commander, who was engaged in a campaign against Port Hudson, the Confederate stronghold on the Mississippi about sixteen miles upstream from Baton Rouge. Banks had been trying to capture the fort since May, and when infantry assaults failed, he settled into a siege. He initially did not give the new arrival a specific task, choosing instead to make general use of his knowledge of artillery and siege tactics. By June 14, however, Stone was in command of the batteries assigned to silence the enemy's guns. Despite dwindling food and other supplies, and almost continuous shelling, the desperate rebels held out until word arrived of the fall of Vicksburg on July 4. This Union victory made Port Hudson superfluous to Confederate strategy, and further resistance would result only in a useless bloodbath. On July 8, a ceasefire was arranged, and once official confirmation of Vicksburg's loss had been received, surrender negotiations began. Banks appointed Stone to lead the Union commission in the talks, which lasted for almost five hours. The Articles of Capitulation, with Stone signing as senior officer, were forwarded to Banks that afternoon. Formal surrender would occur the next day, although many rebels took the advantage to escape. At nine o'clock the next morning, the remaining soldiers laid down their arms.[33]

A short time after the surrender, Banks decamped to New Orleans to bask in the glory of his great victory, but not before ordering Stone to remain and wind things up at Port Hudson. For the next ten days he administered the paroles granted by Banks to the enlisted prisoners and arranged for the transportation of the officers to either Northern prison camps or makeshift New Orleans jails. The paroled rebels took most of their sick and wounded with them, leaving about five hundred of the most serious cases behind to be cared for by the Union medical staff. When the Confederates later asked that these men be transferred to one of their own field hospitals, Stone, on the advice of a captured rebel surgeon, declined. He believed they could receive better treatment at Port Hudson, but promised that "these sick and wounded are paroled and sent to their homes as fast as they become able to journey."[34]

Among his own troops, Stone had to deal with outbreaks of measles and dysentery, which caused "large mortality." The brutal Louisiana summer also took its toll, with many soldiers felled by sunstroke. Low morale was also a problem, particularly among regiments whose terms of enlistment were expiring. These "nine month men" had refused hazardous assignments during the Port Hudson siege, and after the surrender they were given light garrison duty pending shipment home. Nevertheless, many were still restive, Stone commenting that "most of them think of nothing but getting home." Particularly troublesome were the men of the 50th Massachusetts. On July 20, one company balked at doing its assigned tasks. Stone quickly had the entire lot arrested and dispatched to New Orleans and eventually to desolate Ship Island, Mississippi. There the mutineers served a sentence of hard labor for the rest of the war. This object lesson had the desired effect among the remainder of the regiment. There were no more incidents of disobedience. The 50th Massachusetts, however, still had to wait a little longer, as Stone adopted a policy of sending well-behaved units to be discharged first.[35]

Stone faced one issue at Port Hudson that he had not previously encountered: the recruitment of ex-slaves into the Union army. The raising of black regiments had begun under Butler and accelerated with the arrival of Banks. The first black units were composed of free men of color, although later refugees from abandoned plantations and runaways helped fill the ranks. Initially they served as laborers and auxiliaries, but black units fought gallantly at Port Hudson, impressing Banks. In June 1863, he reorganized them into a Corps d'Afrique. Like many white officers, Stone did not have a high opinion of the corps, referring to the black recruits as "worthless soldiers." He strongly opposed the practice employed by General George Andrews, who had charge of the black troops, of raiding small farms that had only a few slaves each to collect them for enrollment in the corps. Writing to Banks, Stone labeled this a "false military move." The small farmers were "the best disposed to the Government I have seen in this State," and served as valuable sources of intelligence as well. Stripping them of their slaves would "make hostile a region now friendly." Stone, nevertheless, did not believe he had the authority to compel Andrews to suspend the conscriptions, but told Banks, "were I in command I would not hesitate to do so, as a measure of military policy."[36]

Despite Stone's complaint, recruitment into the Corps d'Afrique "at the point of a bayonet" continued. By late September, however, officers had begun conducting their own unauthorized impressment campaigns. Some had even taken to raiding government-operated plantations. In forwarding complaints by federal farm managers, Stone again urged his commander to bring the situation under control, warning that "unless some stringent measures are adopted, the oppression of these Negro recruiting officers will become unsupportable by all classes." Banks finally had to act, issuing a proclamation that recruitment of black soldiers could only be done by officers designated by the superintendent of recruiting. Eventually, some twenty-four thousand African Americans from Louisiana served in the Union army.[37]

On July 28, Stone completed his work at Port Hudson and relocated to New Orleans. Three days earlier, Banks had appointed him to be chief of staff for the Department of the Gulf. This was mainly a desk job. He managed a staff of twenty-four officers, and spent his time mainly sending and receiving orders and correspondence for the commanding general, overseeing the work of the quartermaster, commissary, judge advocate, and ordnance departments, gathering and evaluating intelligence from military and civilian sources, and handling transportation and personnel issues.[38]

As chief of staff, Stone served on various military boards and commissions. As soon as he arrived in New Orleans, he found himself again putting his knowledge of ordnance and artillery to work as a member of a special board to develop plans for the defense of the city and the department "in the event of war with a foreign power."[39] Likely this referred to the French, who maintained a naval presence in the Gulf of Mexico. Relations between the United States and France had been strained for some time. During the Butler regime, the French government complained about alleged mistreatment of its nationals in Louisiana, and had a flotilla visit New Orleans to demonstrate its concern. France also had launched its invasion of Mexico, which the United States opposed, and rumors (proved to be unfounded) always seemed to be circulating about a possible Franco-Confederate alliance. In addition, Stone's duties occasionally took him away from the Crescent City, as when he journeyed to Vicksburg to sit on the commission to investigate the August 19 explosion of the steamboat *City of Madison* while being loaded with munitions, in which 150 men lost their lives.[40]

When Banks staged a series of landings along the Texas coast from September through November 1863, Stone and his staff played a key role in keeping the invading forces supplied and equipped. Banks chose to lead part of this campaign personally, leaving Stone to serve as de facto commander in Louisiana. This role brought him into conflict with the newly re-established federal courts system. In November, a steamer called the *Alabama* (no connection to the Confederate sea raider of the same name) had been seized for some reason by the navy and was docked in New Orleans awaiting a prize adjudication in the district court. Stone, in the meantime, badly needed transports to ship supplies and reinforcements to Banks. When he learned of the *Alabama*, he immediately secured the navy's approval for use of the vessel for a month. He then directed that she be loaded and troops put aboard, and ordered her captain to prepare to sail at once. The District Court got wind of this, and the judge, holding that the court and not the navy controlled the *Alabama*, sent a US Marshal and some deputies to put a stop to the loading. Using "violent measures," the marshal and his henchmen evicted all army personnel from the ship and stopped the coaling. In a loud voice he let all present know that he was there to "teach the military authorities to respect the courts." In response, Stone employed his own "prompt and efficient measures" in the form of an infantry company to restrain the marshal and restart the loading and sailing preparations. He later politely informed the US district attorney that the whole matter could have been resolved in five minutes if the marshal had first come to his office, but there was a war on, and the army had to be resupplied and reinforced on schedule, not when the court decided to release ships. Stone concluded with a thinly veiled threat that repetition of such interference in the future could result in the district attorney and marshal being placed in "close custody."[41]

Although Stone may have wished for an assignment that involved a bit more excitement than jousting with obstreperous judges and marshals, his duties did leave him time to enjoy the Crescent City. And in this, he would not be alone. On November 5, after receiving an ecclesiastical dispensation for a mixed union, he married a blond-haired, blue-eyed twenty-two-year-old, Annie Jeannie Stone, who, oddly enough, shared the same last name. The wedding took place in Saint Mary's Church, with Jean Marie Odin, the archbishop of New Orleans, officiating. The new Mrs. Stone hailed from East Feliciana Parish, the site of Port Hudson, and the two may have met when the general was serving there.

The newlyweds would have only a few months together before Stone was once more in the field. Throughout much of 1863, the Lincoln administration had been pressuring Banks to invade Texas for reasons both diplomatic—to send a message to the pesky French in Mexico—and economic—to secure cotton for Northern mills. Banks felt he had complied with these requests through his landings along the Gulf Coast, but these had done almost nothing to dislodge the Confederates from the Lone Star State and were minor annoyances to the French. Even before the amphibious campaign began, Halleck had been strongly urging a move up the Red River, through Alexandria and Shreveport and into Texas from the north. Banks put him off as long as he could, citing sound reasons why such a strategy was impractical, but the general-in-chief would not be denied.

In January 1864, Banks finally agreed to advance up the Red River, but the actual campaign could not begin until after February, when water levels on the Red had risen sufficiently to allow gunboats and transports to navigate. In the meantime, Banks busied himself with political matters relating to the formation of the Free State of Louisiana, while Stone attended to the many logistical details that needed to be addressed in preparation for the invasion. This time, however, he would not remain behind his desk in New Orleans. Stone would be chief of staff of the Department of the Gulf in the field. Ironically, the campaign he was arranging would be into the Confederate department headed by his old West Point classmate, E. Kirby Smith.

Although everything was in readiness, Banks chose not to launch the Red River campaign until after the lavish ceremonies surrounding the inauguration of the new governor of Louisiana on March 4. Stone, Jeannie, and Hettie likely watched the spectacle that included massed bands, an anvil chorus, school choir, and fireworks. Before leaving New Orleans, Stone also penned one last appeal to Lincoln for "some act, some word, some order . . . from the Executive which shall place my name clear of reproach." If the president chose not to do so, the general pledged, nonetheless, to "continue to perform the duties of a soldier wherever I may find myself placed, so long as the Government of my country needs a soldier's services."[42] He received no reply.

Banks sent Stone and a portion of the staff in advance by boat to Alexandria, where they arrived on March 20. In Alexandria they joined the division of Brigadier General A. J. Smith, on loan from William T. Sherman in Mississippi, along with the gunboats and transports under the command of Commodore David Dixon Porter, and waited for the rest of the expedition. While in Alexandria, Stone busied himself with issues of forage and supply, keeping track as best he could of enemy troop movements. Ever mindful of the value of good intelligence, he began recruiting a detachment of "Western Louisiana Scouts" to provide information on the region to Banks. Divisions under Major General William B. Franklin, whom Stone had commanded during the "siege" of Washington in 1861, did not begin arriving until March 25, and Banks himself did not show up until the next day. When his command finally gathered, Banks pushed upriver to the hamlet of Grand Ecore, where he halted to ponder the best route to Shreveport.[43]

Aboard Banks's headquarters steamer the *Black Hawk* at Grand Ecore, Stone met with veteran Red River pilot Wellington Withenbury. The two exchanged maps, which in a number of cases showed different locations, features, and roads. Banks joined in on the conversation, and, with Stone, asked the pilot about the best way to Shreveport. Withenbury responded that the better road was up the east side of the Red River, but it would take two or three days longer than the one through Pleasant Hill and Mansfield on the west side. The latter road, however, would take the column away from the protection of the navy's gunboats. In the end, Banks, probably with Stone's concurrence, decided that time was precious and opted for the Pleasant Hill–Mansfield route. Banks had to return Sherman's division by April 10, and the new commanding general of all Union armies, Ulysses Grant, expected him to be in Shreveport by the end of the month. He felt so pressed that he did not even send cavalry ahead to scout the countryside before beginning the march. That would take too much time. Stone, who appreciated the value of accurate military information, likely did not agree.[44]

By the time the army commenced leaving Grand Ecore on April 6, the relationship between Banks and Stone had definitely deteriorated. Heretofore, Stone's work as the chief of staff of the Department of the Gulf had been generally satisfactory. The two men had gotten along well enough at Port Hudson, where the commanding general made use of Stone's expertise, and in the fall when Banks was away on the Texas coast. Once they were in daily contact in New Orleans, and particularly during what was shaping up to be a stressful campaign, however, things changed. At one point, Stone responded to criticisms from Banks by submitting a letter of resignation. Banks did not act on it, but came to regard him as "a weak man." To be successful, a commander must have trust in his chief of staff, and the chief must have confidence in his commander. In early April, both trust and confidence were in short supply along the Red River.[45]

The march toward Shreveport proved particularly taxing. One soldier described the route through thick pine forests as "a howling wilderness." Rain turned the red clay road slick and slowed progress to a crawl. The line of march upon which Banks insisted, with infantry and cavalry units and wagons bunching up and then extending like an accordion, also impeded the advance. As the army approached Mansfield on April 8, it collided with a Confederate force under Major General Richard Taylor at Sabine Crossroads. When the battle began, Stone was riding with the forward cavalry column. He remained at the front to reconnoiter rebel infantry and artillery positions, and at one point led a small detachment of foot soldiers in a gallant, but unsuccessful, attempt to shore up the Union's collapsing left flank. But the day was lost. Stone tried in vain to rally the frightened soldiers. Even Banks, who also was at the front, could not stem the Yankee retreat. Only sundown ended the disorderly flight. Late in the day when Taylor received instructions from his commander not to engage the enemy, he could only reply, "Too late sir. The battle is won."[46]

Fighting resumed the next day at the village of Pleasant Hill. Banks attempted to reform his defeated army and adopted a defensive strategy. His deployment, however, proved to be haphazard. He left some units isolated and others unsupported, certainly not in the manner they would have been positioned by a trained officer. Stone evidently had given up trying to advise Banks or had his advice ignored. At any rate, before the battle at Pleasant Hill began, Banks was seen walking about conversing with the troops, while Stone "was sitting on a rail smoking cigarettes, and apparently more interested in the puffs of smoke that curled around him than in the noise and bustle around him." Rather than going over plans and issuing orders, the two men seemed to be ignoring one another, despite the fact that a desperate struggle was about to unfold.[47]

Once the Confederates advanced and fighting started, Stone again rode to the front where he transmitted orders and directed troop movements, no easy task given Banks's disjointed placement of his troops. At one point he was called upon to observe how Colonel William Shaw arranged his brigade in the face of a superior number of the enemy. He approved and ordered the position be held at all costs. He promised to send reinforcements, although none arrived before Shaw was told to retire. Stone did, however, see some direct action. About four P.M., as a portion of the federal line was in danger of being overrun, he led his staff and a part of A. J. Smith's division in what an observer called "a most magnificent charge," which, "with terrible effect," stopped the rebels and forced them into retreat.[48]

Despite Stone's heroics and the dogged resistance of the Union troops in holding back Taylor's advance, Banks ordered a retreat under cover of darkness to the relative safety of Grand Ecore. His men, while glad to be alive after two days of almost continuous fighting, were dispirited. Many felt they could have whipped the rebels, but had been betrayed by poor generalship on the part of their commander. They sang derisively about how "we all skedaddled to Grand Ecore," and shouted, "Napoleon P. Banks!" Some took to calling him "Mister Banks," a reference to his lack of military training and experience. In addition to dealing with the disaffection of his soldiers, Banks had to spin the reversal at Sabine Crossroads and his decision to pull back after stopping the Confederates at Pleasant Hill to the Union high command. He told Grant that he had actually won both engagements and was planning to take the offensive again, albeit by a different route. Clearly this was not going to happen, at least not anytime soon, and Grant knew it. He had his own agenda, and had warned Banks that if he could not be in Shreveport by April 30 to return to New Orleans and get ready to assist in the capture of Mobile, which is what Grant and Stone had wanted to do in the first place.[49]

As part of his campaign to shift responsibility for what Sherman called "one damn blunder from beginning to end" away from himself, Banks began making changes to his command. The first to go was his chief of staff. He tried to blame Stone and other generals for the defeat. On April 16, Banks received instructions from Washington and issued Field Order Number 21, relieving Stone from all duties with the Department of the Gulf and ordering him to New Orleans and then to Cairo, Illinois, to await further instructions. He did not like Stone, and on who better to pin the blame for the muddle on the Red River than the scapegoat of Ball's Bluff? Besides, Stone had married a Southern woman whose brother was serving in the rebel army, raising old suspicions that the general might, indeed, be soft on the Confederacy.[50]

Stone might have been expecting his firing since he declined to protest and dutifully packed up and with his family began his journey to Cairo. Perhaps he was secretly relieved to be out from under Banks's thumb, but he was certainly shocked when the letter from the War Department, dated April 4, was turned over to him. He saw that it was addressed to Colonel C. P. Stone! He had been mustered out of the volunteers and demoted from brigadier general to his regular army rank, and was once again an officer without an assignment.[51]

Technically, Stone had been placed on leave, but an old pattern began to emerge. He contacted the adjutant general upon arriving in Illinois but got no reply. As spring turned to summer, his entreaties became more insistent, even threatening to resign if he could not be restored to active duty. Again silence. Finally, he informed Washington that his wife had become ill due to unhealthful conditions at Cairo, and, if his services were not needed, he wished to accompany her to Massachusetts where the climate might aid her recovery. There was no answer, so Stone stayed in Cairo, and Jeannie got better.[52]

Stone might have sat out the remainder of the war on leave in Illinois had it not been for his old acquaintance and fellow West Point cadet, Grant. Grant had approved the removal of Stone from Banks's command, possibly to separate him from the debacle that had become the Red River expedition. Stanton, after demoting Stone, had tried to poison Grant's opinion of him by falsely representing him as having been in command during the reverses in Louisiana. Grant paid this no mind, and probably at his behest, on August 13 the adjutant general's office directed the colonel, who was on leave in Massachusetts, to report "in person and without delay" to the Army of the Potomac before Petersburg and Richmond. A week later Stone arrived at the headquarters of Major General Gouverneur K. Warren and was assigned to command the 1st Brigade, 2nd Division of the 5th Corps.[53]

Stone's tenure in Virginia with the 5th Corps, however, turned out to be brief. A biographer noted that by this time he was "broken in heart and health." Although he had assumed temporary command of his division by early September, rumors circulated that he was showing signs of nervous exhaustion. He also may have contracted a case of typhoid. Soon, Stone came to realize he could not lead his troops effectively, and he knew that if his division or the 5th Corps suffered any reverses, Stanton would be quick to put the blame on him. His commanding officer was disappointed to see him go, but Stone had had enough. On September 13, just three days after the birth of his daughter Fanny, he resigned his commission, leaving the United States Army for the last time.[54]

STONE PASHA

Stone left the Army of the Potomac with his honor, if not his physical and emotional health, intact. Along with Jeannie and Hettie, he moved to Massachusetts to begin his recovery. Amid familiar and friendly surroundings, and away from the stresses of his unfortunate army career, Stone progressed rapidly. By the spring of 1865, he was planning a visit to Mexico City. The reason for this journey is not exactly clear, but he was likely still pursuing some type of payment under the Jecker Contract. Now out of the military, Stone needed money. He may have believed that with Maximilian in power the Mexican government would reverse the decision of the previous administration cancelling the agreement with Jecker and look kindly on claims filed against it.[1]

While aboard the steamer headed to Mexico, Stone bumped into an old acquaintance from California, Dr. William McKendree Gwin. Back in the 1850s, Gwin had been one of the most prominent political figures in the Golden State. Almost twenty years Stone's senior, he was a Southerner who came west during the Gold Rush. He organized and led the "Chivalry," or pro-Southern, faction that dominated California's Democratic Party. Elected as one of the state's first two United States senators, Gwin served two terms. Once the Civil War began, he returned to Mississippi, but when his plantation was destroyed he moved to France. There he promoted the settlement of slaveholders in French-controlled Sonora. It was in pursuit of this project that he journeyed to Mexico and became involved with Stone. After reminiscing about the old days in California, the two men probably commiserated on their recent misfortunes and shared plans for the future. Gwin succeeded in convincing Stone to participate in the colonization scheme, while, for his part, Stone provided expertise on the geography, economy, and politics of northern Mexico. When the government of Maximilian refused Gwin's proposal in the summer of 1865, fearing that the immigrants might alienate the land, and that the settlement of former Confederates would hinder the empire's bid for diplomatic recognition from the United States, the ex-senator departed. Stone remained in Mexico City into the fall, still pursuing his original mission of getting paid for his claims. He did not realize a single *peso*.[2]

Upon returning empty-handed to the United States, Stone was under even more pressure to find a source of income to support his family. Military service was not an option, so he turned to engineering. In late 1865, a consortium of Northern businessmen headed by shipping magnate William H. Aspinwall acquired controlling interest in a group of coal mines in Goochland County near Richmond, Virginia. The following January they incorporated the Dover Company to operate the mines. Stone became a director of the firm and its superintendent and chief engineer. He also found employment as a consulting engineer for the James River and Kanawha Canal Company, a venture that sought to link the Tidewater region of Virginia with the Ohio River by means of a series of locks.[3]

The Dover mines had been producing coal since the mid-eighteenth century and had supplied the Confederacy's Tredegar Iron Works. With the coming of peace, the mines needed modernization, and over the next three years, Stone sank new shafts and installed steam pumps to dewater existing excavations. Under his direction the company established a farm to feed the miners as well as a block of houses and a store for the workers. In addition, Stone built a large frame Victorian home for himself and his family, which increased with the birth of a son, John, in September 1869.[4]

All of the improvements to the Dover mines undertaken by Stone cost money, and, despite his hard work, the firm never turned a profit during the postwar years. Instead it accumulated a substantial debt as the coal it did produce turned out to be of inferior quality. By the end of 1869, Stone had been removed as a director and let go. The company failed the following year. Nor was the canal venture any more successful. Although Stone's analysis found the project to be feasible from an engineering standpoint and a worthwhile undertaking, in the 1860s railroads superseded canals as the nation's primary means of transportation. Trains were the shiny darlings of Wall Street, and the James River and Kanawha Canal Company could not obtain financing. It also failed.[5]

After leaving Dover Coal, Stone found himself once again "financially embarrassed." He sought work, likely as an engineer, although now he could list mining on his resume, which was becoming fairly long on experience if somewhat short on successes. When he did find employment, it came from a totally unexpected quarter—an old friend from California.[6]

In 1869, William Tecumseh Sherman, now commanding general of the United States Army, was contacted by Thaddeus Mott, an American soldier of fortune who had seen action in Italy and Mexico and in the Civil War. At the time, Mott was serving as a major general and aide-de-camp to His Highness, Ismail, khedive of Egypt, also known as "Ismail the Magnificent." The khedive, like his pharonic predecessors, had a grand vision. First, he wished to transform Egypt from a province of the declining Ottoman Empire into an independent nation. He also sought to modernize and westernize the country and extend its boundaries and influence into equatorial Africa. To do this he needed an army trained and equipped along European lines. But he felt he could not trust the Europeans. They were beginning to expand their own empires and might just gobble up his along the way. He did not want to trade a master in Constantinople for one in London, Paris, or even Saint Petersburg. When Mott suggested to him that American officers could be of benefit to the Egyptian army, the khedive, having recently dismissed his French military advisors following a dispute with France over the Suez Canal, became intrigued. The United States had few interests in the Middle East or Africa, but what it did have was an overabundance of experienced soldiers, many of whom were either unemployed or underemployed. Ismail ordered Mott to go and recruit.[7]

Mott turned to Sherman for recommendations of Civil War veterans, both Union and Confederate, who might be interested in serving the khedive. At the top of the list was Sherman's Gold Rush business associate, Charles P. Stone. Heavily in debt and in need of a job, Stone readily accepted Mott's offer of a five-year contract in Egypt. In the spring of 1870 he accompanied a contingent of officers on the three-week journey to the ancient port of Alexandria. They were not the first American soldiers to enter Ismail's service. Mott had brought ex-rebel generals William Wing Loring and Henry Hopkins Sibley with him when he returned from his recruiting mission. A few others followed, but upon his arrival, Stone quickly became recognized as the leader of the American officer corps in Egypt. After all, he had been, for a while at least, a general on the winning side of the Civil War. His hair was thinner and turning white, but he still stood ramrod straight and possessed the accomplished manner of one used to command. Stone's experiences abroad also gave him a cosmopolitan outlook. Unlike many of his fellow expatriates, he would be neither repulsed nor intimidated by the strange land in which he found himself.

Stone and the other new arrivals did not tarry long in Alexandria, but quickly made their way by train to Cairo. There they secured accommodations and uniforms and had their first audience with the khedive. While some of the Americans may have expected to encounter an Oriental potentate out of the Arabian Nights clad in robes and a turban, Ismail appeared in dark trousers, a high-collared dark coat festooned with decorations, a white shirt and waistcoat with cravat, and a green sash across his expansive chest. He had a finely clipped beard. The only exotic touch was the traditional cylindrical red fez with black tassel. Though born in Cairo, he was thoroughly westernized, having been educated in Paris and traveled extensively in Europe before being proclaimed khedive in 1863. Charismatic in his way, Ismail was not overbearing but came across as a man accustomed to being listened to when he spoke. Loring probably best captured the singular appearance of Egypt's master:

When in repose and his eye is partly shut, no man has a more sphinx-like expression; but the strongly-marked face conceals behind it a constant thought and indicates that the cares of state weigh heavily upon him. In his hours of ease his conversation is very agreeable. Speaking French slowly and deliberately, with a finely modulated voice and a countenance lit up with the characteristic smile of his family, he gives one the impression that he would make a boon companion.[8]

Many of the American officers took a liking to Ismail, but none more so than Stone. He was deeply impressed by His Highness and his progressive ideas for Egypt. After almost three years there, he wrote that the country was struggling to be "more civilized and 'liberal' " than the so-called civilized nations of the world. Under Ismail, it sought to "grow strong and well regulated."[9] Stone would stand at the side of the khedive steadfastly until the end of his reign and beyond. For his part, Ismail was impressed with Stone as well. When Mott abruptly left the royal entourage in August 1870, Stone assumed the role of trusted advisor, helping to select more Americans to serve in Egypt. He became part of the khedive's inner circle, and, as one American recalled, "only one American officer could get an interview [with Ismail] whenever he wished it, and that one was General Stone."[10] Stone's influence expanded when Ismail made him chief of staff of the Egyptian army, with the rank of lieutenant general, and extended him the title of Ferik Pasha. As additional recompense, Stone and his family were given free use of a palace in one of Cairo's most fashionable neighborhoods.[11]

A number of his colleagues left their families back in the United States but Stone brought his with him to Egypt. For Jeannie, Hettie, whom they called "sister," Fanny, and "Johnny Boy," living and growing up in a palace complete with a retinue of servants, must have seemed like a dream. Like the offspring of other Americans in Cairo, they attended a "European" school where they learned French, mathematics, and classical studies. Unlike their playmates, however, the young Stones explored as much of their exotic surroundings as mother and father would allow, and even mastered some Arabic.[12]

Jeannie divided her time between trying to maintain a proper Victorian household and a raft of social duties as the leader of the families of the American military community in Cairo. In both pursuits, she had the able assistance of her stepdaughter, Hettie, who was becoming a young lady. The Stone's home became a gathering place for the American officers and their families. As the wife of the chief of staff, Jeannie had the responsibility for greeting newcomers and helping them become acclimated to their exotic surroundings, hosting birthday parties and celebrations at Christmas, New Year's, and American national holidays, and organizing excursions to the pyramids and monuments.[13] She also would have accompanied Stone to lavish entertainments at the Gezirah and Abdin palaces to welcome distinguished visitors and commemorate accomplishments of His Royal Highness. An observer at the court later described one such evening:

> At nine o'clock the company assembled in the new wing of the palace, where the Khedive received the guests with his usual urbanity, conversing with ladies and gentlemen, previously known to him, with much affability. About 150 invitations outside of his immediate court circle had been issued, and all the nations of Europe were represented by richly dressed women, and men in the somber suits which the nineteenth century renders *de regle* for full dress. About an hour was occupied with this reception business, and then the Khedive, with a lady on his arm, followed by the young princes, each escorting a lady, led the way into a long saloon prepared as a concert-room; where a concert was given by the best singers from the opera troupe, male and female. When this was over, the company moved back into the other apartments, of which there was a long suite. The chairs were removed from the concert-room, which was converted into a ball-room. The band struck up, and dancing began, which was kept up until long after midnight, when the doors to the supper-room were thrown open, and the cuisine vied with Terpsichore for a time. It was a very curious and picturesque sight, to see the strange blending of nationalities and costumes, Western and Eastern. The Khedive's officials and court were in gorgeous uniforms, their breasts sparkling with decorations. The Khedive himself does not possess or flourish the fantastic toe; his weight, both of person and character, preventing. The ball was kept up with great animation until the "wee sma' hours," the Khedive manfully holding his ground until the latest revelers had departed.[14]

Jeannie did take some time away from her social duties in 1871 for the birth of a daughter. She and the general named her, appropriately, Egypta, and since Egypta was born on November 1, All Saints' Day, she was given the middle name, Todassantas. Informally, the family called the baby "Todas."

The khedive had made Stone his army's chief of staff, but in 1870, the staff did not exist. Ismail's grandfather, Mohammed Ali, had developed a functioning general staff, but in the decades following his death in 1849, it had been allowed to disintegrate. It now became Stone's responsibility to put it back together.

Since neither the khedive nor his minister of war really understood the duties of the general staff, Stone was given a free hand in fashioning it. He explained to them why such an organization was needed: "The body has a head, arms, legs, but there are no nerves."[15] The general staff would be the army's central nervous system. A subordinate later described how he went about his task:

> When the quiet and politic American, General Stone, was made Chief-of-Staff, of the army, he did not announce himself with a blare of trumpets; in fact he kept in the background, and few of the high officials were aware of the fact that a new power had arisen among them which had to be reckoned with. Stone worked silently and unobtrusively, but he had the tremendous leverage of the Khedive's power to help him.[16]

American officers formed the basis of the general staff, although educated Egyptians took positions on it as well. Their headquarters was in the former harem wing of the medieval palace overlooking Cairo known as the Citadel. Over the next few years, Stone and his officers attempted to install departments for quartermaster, paymaster, commissary, ordnance, mapping, engineering, and intelligence. Stone set up in the Citadel a small military museum and a library with standard texts on tactics and strategy along with journals from around the world. He even succeeded in establishing a military printing office for the publication of maps and reports. Eventually, the government's Department of Public Works would come under the authority of the general staff as well. Stone realized that a trained officer corps would be vital to the success of the army. He convinced the khedive to found a military staff college for which he developed the curriculum, and his officers served as instructors.[17]

The results of the efforts of Stone and the Americans were mixed. Not all of the bureaus became fully functional or achieved their goals. Quartermaster, paymaster, and commissary duties in particular would continue to be handled at the unit level, a very cumbersome and inefficient approach. The lack of unified quartermaster and commissary services also would contribute noticeably to Egypt's military reverses later in the decade. In addition, while the staff college turned out first-rate officers, graduates often found themselves distrusted by their more traditional commanders. They were frequently passed over for promotions in favor of candidates who remained on duty thereby gaining a year's seniority. On the other hand, the mapping and surveying activities undertaken by the general staff achieved substantial successes.[18]

Another successful aspect of Stone's program was the education of the army's enlisted ranks, almost all of whom had been conscripted from the fellahin, or peasantry, and were illiterate. He established a basic literacy program, and eventually some seventy five percent of the enlisted personnel would learn to read and write at a level where they could fill out leave forms and applications for promotion. When Stone learned that the sons of corporals and sergeants tried to attend classes with their fathers, he extended the literacy program to include them as well. He proudly wrote to an acquaintance, "It is astonishing to those who look upon the Egyptian fellah in the fields to see him after a few months service in the army, reading and writing.[19]

Stone's educational reforms coincided with Ismail's vision for his people, but it also had a practical military side. It advanced the general's goal of loosening the grip of the civilian Coptic clerks on army operations. The clerks handled all paperwork at company and battalion levels and were often inefficient and corrupt. By developing a cadre of literate noncommissioned officers who could assume the clerical functions of the Coptics, he hoped to eliminate what he perceived to be a major obstacle on the road to creating a modern Egyptian army.[20]

A greater problem for Stone in trying to modernize Egypt's military lay in its methods of command. Senior officers, most of whom were pashas, had total control of all aspects of their units—supply, rations, ordnance, pay, and promotions. This had been the way the army had functioned since the days of Mohammed Ali, and the system was well entrenched. Try as he might, despite his influence with the khedive, Stone could not break it down. Another frustration Stone encountered was the monopoly that Turks and Circassians held on positions of high rank. No matter how talented, a native Egyptian officer could not advance beyond the rank of lieutenant colonel, and only a few made it that far. Stone argued before the khedive against this throughout his tenure in Egypt, but like the pasha system, it was too ingrained in the country's military establishment. Pashas and Turco-Circassians were irritants Stone would just have to accept. The acerbic but perceptive ex-Union officer William McEntyre Dye recounted as much in his book, *Moslem Egypt and Christian Abyssinia*, "A stranger might have supposed . . . that [Stone] was satisfied with these things as they were, but I thought it merely a reluctant assent given to what was considered the inevitable."[21]

As chief of staff, Stone did have authority over the rearming of Egypt's military. By the 1870s, its small arms and artillery were largely obsolete and in need of replacement. Stone was able to put his knowledge of American and European arms manufacturers to use in making recommendations to the khedive. He promoted E. Remington and Sons of New York to be the principal source of rifles, carbines, and revolvers, and the Remington rolling block rifle became the standard weapon of the Egyptian infantryman until reorganization of the army by the British in the 1880s. Although Stone would like to have directed artillery purchases toward an American company, in the end Krupp became the supplier of coastal defense guns and other cannon, mostly surplus from the Franco-Prussian War. Under his direction, the Egyptian government also established factories to produce some small arms and ammunition.[22]

Stone's high rank and position within Ismail's inner circle took him abroad as the khedive's special representative at international conferences and events. He also organized and oversaw the installation of the Egyptian sections at exhibitions in Vienna in 1873, Philadelphia in 1876, and Paris in 1878. His cosmopolitan outlook, polished manners, and fluency in French, Spanish, German, and Arabic made him ideal for such assignments.

Eventually some fifty Americans from both the North and South served in Egypt, though not all at the same time. By and large, they held staff positions and did not directly command native troops. Although Stone probably believed most of his recruits to be superior field officers, wide differences in temperament, culture, language, and religion between the Americans and the Egyptians limited their duties. Only once did the khedive entrust an American with command of a significant body of troops, and that officer was Stone. When the president of the Suez Canal Company, Ferdinand de Lesseps, threatened to use armed force to resist a reduction in tariffs ordered by the sultan in Constantinople, Ismail directed Stone to take a battalion to the canal and sort things out. He carried out the mission and returned to Cairo without a shot being fired.[23]

Service on the general staff did not mean that Stone's Americans remained behind their desks at the Citadel. Probably the most successful aspect of the American presence in Egypt was the exploratory work done in furthering Ismail's ambitions in Equatorial Africa. For the khedive, the words of Egyptologist Mariette Bey rang very true: "History should say that wherever flows the Nile, there her [Egypt's] rights and her dominion should extend."[24] Ismail's grandfather, Mohammed Ali, had conquered the Sudan, and now he was poised to extend his empire until Egypt controlled Africa's northeast quarter. From 1870 to 1876, officers of the general staff conducted some thirty expeditions and reconnaissances, mapping Egypt and the Sudan and exploring regions to the south. As chief of staff, Stone authorized, planned, and organized most of these adventures, and his office issued the maps and reports generated by the surveyors and explorers. Many of these would be presented at gatherings of the Khedival Society of Geography in Cairo, of which he was a founder and later president.[25]

Of Stone's American officers who participated in African exploration, none had a more exciting time than did Charles Chaille-Long. A Union soldier during the Civil War who rose from private to captain, he held the rank of lieutenant colonel in the Egyptian army. His service had been unexceptional until 1874 when, at Stone's likely urging, the khedive recommended him to Charles "Chinese" Gordon, his newly appointed governor of Equatoria, a province with headquarters at Gondokoro some seven hundred fifty miles south of Khartoum. The colonel would serve as the governor's chief of staff. In later years, Chaille-Long claimed that before leaving to join Gordon, he had a secret meeting with Ismail to receive confidential instructions to the effect that his main purpose in going south would be to protect Egypt's interests, particularly against the British. He was to meet Gordon in Gondokoro, but must proceed quickly on his own to make a treaty with King M'Tesa in Uganda before the American explorer Henry Stanley, leading a British expedition, could get there. While little evidence exists to back up this story other than Chaille-Long's assertions, if such a meeting did occur, Stone, as Ismail's chief military advisor and confidant, would certainly have been in on it. As it was, Stone played a key role in planning Chaille-Long's expedition, arranging equipment, supplies, and transport. Whether Chaille-Long's story of the secret meeting was true or a fabrication, the khedive and Stone did not quite trust Gordon, unsure what to make of the enigmatic and eccentric Englishman. Where did his loyalty really lie? In Chaille-Long they felt they had someone they could count on in his inner circle. Stone likely reminded the colonel prior to his departure that while he may be Gordon's chief of staff, he ultimately reported to him and to the khedive.[26]

Under Gordon's immediate direction, Chaille-Long undertook extensive and grueling treks into Central Africa, in the process discovering Lake Kyoga (which he called Lake Ibrahim), part of the system of lakes and rivers on the upper Nile. He succeeded in obtaining a treaty of friendship with King M'Tesa. Cairo interpreted the agreement to mean that Uganda became a vassalage, while M'Tesa did not see it quite that way. In addition, in one engagement, Chaille-Long claimed to have fought off about seven hundred "enemies of the Khedive." He returned to Gondokoro in March 1875, and to Cairo two months later, bringing with him—in addition to the treaty—notes, journals, artifacts, specimens, a daughter of M'Tesa, and a pygmy named Tiki Tiki, who entertained the royal court. He reported his activities to Ismail and Stone, enjoyed a public reception "in the nature of a triumph," and spoke before the Khedival Geographical Society about his exploits. Stone ordered Chaille-Long promoted to full colonel and then gave him leave to recuperate in France. He would later be recalled by the general for further adventures.[27]

In addition to organizing expeditions, coordinating with officials such as Gordon, managing the general staff and serving as confidant to the khedive, Stone bore the at times frustrating responsibility for maintaining the order and discipline of his American soldiers of fortune. This involved mediating disputes with their employer and among each other. In a way, he was like a schoolmaster trying to convince his rowdy pupils to play well together and get along with the other children. It was a particularly sensitive job, since less than a decade earlier these men had been pitted against one another in a bitter conflict. Insults between ex-Yankees and rebels were common, although infrequently acted upon. In addition, while they may have admired Ismail and enjoyed his entertainments, the Americans, by and large, did not speak to or associate with their Egyptian counterparts except on a strictly professional level. Many also brought their xenophobic ideas with them and had little interest in Egypt, outside of the antiquities, and little understanding of its people, culture, and religion.[28]

Conflicts between the Americans and Egyptians appear frequently in accounts written by the officers of their time in the Middle East. These range from battery on unfortunate beggars, vendors, and camel drivers to more serious incidents such as the encounter between a hotheaded young Major James Morris Morgan and an Egyptian official:

> It appears that on one occasion General Loring, along with two of his staff, Morgan and Chaille-Long, strolled into the *Buvette* of the *Theatre Francais*, Cairo, during the interval of a gala performance which was being given there that night. Ali Bey, prefect of police, was there, attired in the gorgeous uniform of an officer of the Khedive's staff. As he entered, the Egyptian saluted General Loring, and addressing Major Morgan said: "Give me a glass of water." Morgan turned blue, filled his own glass carefully and deliberately, and rising with calm dignity, dashed the contents full in the Prefect's face. Before the surprised Bey could recover from the shock, he had his cheeks slapped by the imperturbable Morgan so energetically that the Egyptian was almost swept out of the door.[29]

While in this case, Ismail himself smoothed over the ruffled feathers, more often than not it was up to Stone to make sure such occurrences, along with incidents of drunkenness among his officers and their reluctance to pay their debts to local innkeepers and merchants, did not endanger the American position in Egypt.

Visits to Egypt by dignitaries, particularly acquaintances from his Civil War days, brought Stone a welcome respite from the stresses of his job. In 1872, William Tecumseh Sherman arrived in Alexandria. Initially, he was dismayed by the poverty and squalor he saw around him, and felt insulted when Egyptian officials deferred to his aide, Lieutenant Frederick Grant, son of the general and serving president, considering him a prince of the royal blood. His mood improved, however, when he met with Loring and his old friend Stone. So impressed was Sherman with the work the chief of staff was doing that he agreed to continue recommending soldiers for the khedive's army. Two years after Sherman's visit, George B. McClellan swept through Egypt, also enjoying a reunion with Stone and other Civil War veterans. Of the meeting he would write in his diary, "Stone was kind and agreeable as he could be. He looks very well and is happy."[30]

The most important dignitary to travel to Egypt came near the end of the American military experience there. In 1877, ex-president Grant and his wife, Julia, embarked on an around the world journey—the grandest grand tour. After eight months in Europe, the party sailed to Alexandria. To the Egyptians, their arrival on January 5, 1878, amounted to a visit from the "King and Queen of America." As the khedive's principal military advisor and the army's chief of staff, as well as someone who knew Grant, Stone was expected to play an important role in the ceremonies and events surrounding the visit. Ismail furnished a special train to transport the Grants and their entourage from Alexandria to Cairo. As the train pulled into the Cairo station, Stone was there to greet them. Grant recognized the general right away on the crowded platform and remarked, "There's Stone who must have been dyeing his hair to make it so white." Stone was the first to board the car and presented the khedive's personal representative to the ex-president and first lady. They all then boarded carriages for the drive to the palace in which the party would be staying. Grant rode with his old friend Loring, while Stone, whom Julia described as "gallant and brave," rode with her in the second carriage.[31]

The next day, Grant called formally on the khedive, who followed suit by calling on the Grants. Ismail spoke in French and Grant in English. Through a translator, the ex-president talked glowingly of Stone, reminding His Highness that they were old friends from Mexican War days. He believed Stone to be a superior soldier and was pleased to see him as the khedive's chief of staff. Ismail replied that he was pleased with Stone, in particular with the way in which he organized troops. Later Grant would comment also that at the beginning of the Civil War Stone was regarded as one of the Union's top generals, but that he "had been the most unfortunate man he had ever known."[32]

That evening the United States consul general in Cairo hosted a dinner party for the Grants to which a number of the American officers and their wives were invited. Among the guests were Jeannie and Hettie Stone. Effusive toasts were offered to the khedive, who could not attend because the royal household was in formal mourning over the death of the king of Italy, and to General and Mrs. Grant. Stone then made one of the night's principal speeches in praise of both His Highness and the ex-president.[33]

Grant felt more at ease on a visit to the Citadel. He once again was among soldiers, many of whom he knew, and some he had fought against. The younger officers and cadets stood at attention, in awe of the American war hero. Stone was there to escort him through the venerable fortress palace, and took pride in showing the former commander of the Union armies the various offices and bureaus of the general staff. At Grant's request, the Stones accompanied the party on a tour of Cairo and an expedition to the pyramids and the sphinx. On January 16, the Stones were present to see the Grants off on a steamboat voyage up the Nile. Before departing, Stone and Loring sat on the veranda deck reminiscing with Grant about their experiences in the Mexican and Civil wars. When the travelers returned several days later, Stone journeyed south to meet them at Memphis, where he provided a tour if its ancient ruins and monuments. Once back in Cairo, the Grants enjoyed another round of receptions and dinners, one of which was hosted by the Stones, before leaving for the Holy Land, Greece, and beyond.[34]

When, a year later, the Grants again passed through Alexandria en route to India, Stone once more was there as Ismail's representative. He escorted them to Suez, where they caught a steamer to the east. By that time, however, everything had changed for Stone, his American officers, and the khedive.[35]

The first hint that something was wrong came in the mid-1870s. Years of Ismail's overspending and borrowing from the European powers to support public projects and private indulgences had begun to crack the Egyptian treasury. The payment of salaries for government employees, including Stone and the rest of the army, became problematical. Since no central paymaster's office existed, paydays had always been irregular, but now they were fewer and farther between. This set the Americans to grumbling, since many had families to support either in Egypt or back home. Nevertheless, Stone and most of his staff remained at their posts, and it would not be long before they would become enmeshed in the khedive's conflict with his southern neighbor, Abyssinia.[36]

For some time Egypt and Abyssinia had disputed territories along their mutual border. In 1875, Ismail decided it was time to enforce his country's claims. He also sought to dominate the profitable caravan routes from Uganda. To accomplish these objectives the khedive sent out four expeditions. The first went to Massawa, a port on the Eritrean coast of the Red Sea controlled by Egypt, and then on to the mouth of the Juba River. The force would be under an ex-Royal Navy officer, Henry McKillop, who commanded the naval portion of the operation, and Chaille-Long, who would lead troops inland along the Juba, forcing an opening to Uganda. The second was to engage the Abyssinians in what Ismail described as "a defense of the Egyptian frontier." It would be led by an ex-Danish army officer, Lieutenant Colonel Soren Adolph Arendrup. Arendrup served as Stone's first secretary on the general staff, where he had a reputation as an able administrator and somewhat of an expert on artillery. Although Arendrup possessed no experience in leading troops into a combat situation, Stone liked and trusted him, and he approved of the War Ministry's decision to place the Dane in command. As in the case of another inexperienced subordinate that Stone trusted, this would turn out to be a bad decision.[37]

The other two campaigns in East Africa would be led by the flamboyant Werner Munzinger, a linguist and ex-French and British consul who was now with the Egyptian army, and Brigadier General Mohamed Rauf Pasha, a veteran soldier. Munzinger was to seize the important Abyssinian salt source at Aswa inland from the southern end of the Red Sea, and subdue the Showa Kingdom in the central highlands. Rauf, in the meantime, would march on the important Abyssinian trading city of Harar.[38]

Stone had promoted this ambitious four-pronged offensive as more than just a means of settling a border squabble. He saw it as extending Cairo's supremacy over the East African coastal region, punishing the Abyssinian King Yohannis, and surrounding him with territory controlled by Egypt. Stone, Gordon, and Munzinger lobbied the khedive to adopt this strategy and presented him with plans for its execution. Ismail, who wished to check European, particularly British, influence in the region, agreed. After all, his governor in Massawa had assured him that "good maps, some intelligent officers and three or four thousand well armed men" would be sufficient to ensure victory. These were just the ingredients that Stone and his Americans had been preparing over the past five years.[39]

By 1875, the Egyptian army did appear formidable, at least on paper and on the parade ground. It possessed the accouterments of a modern fighting force—howitzers, Gatling guns, and new Remington Rolling Block rifles. The regiments had mastered the manual of arms and close order drill, and marched in step. Ismail was now ready to put his troops to the test.

The campaign began with high hopes in September 1875, but soon everything went wrong. Arendrup advanced quickly from his base at Massawa. Some of his subordinates requested that Stone send orders halting or at least slowing Arendrup's movement until information could be obtained regarding the location and strength of the Abyssinians. The chief of staff, however, had little regard for his opponents and did not heed the warnings. Arendrup also made the mistake of dividing his force, detaching men to occupy various locations along his route of march. At a pass called Gundet in the Abyssinian highlands, a large enemy force attacked and destroyed the main column, by this time consisting of about eight hundred Egyptian and Sudanese soldiers. Only a handful survived the half-hour-long battle to make it back to Massawa. Arendrup was dead.[40]

Munzinger met a similar fate. His intention of marching with about four hundred Sudanese troops to the salt plains of Aswa had been supported by Stone. But the expedition was poorly planned and executed. Scorching heat and a lack of rations and transport slowed the advance. On November 14, a surprise attack by fierce Somali tribesmen overran the Sudanese camp. Munzinger and his wife, whom he had brought along, died in the assault. Less than half of the command made it back to the relative safety of the coast.[41]

McKillop and Chaille-Long did not fare much better. They landed a force of three companies near the mouth of the Juba River, but were immediately told by the local inhabitants to leave. Chaille-Long refused, established a fort, and began exploring the Juba River. In the meantime, the sultan of Zanzibar, who nominally controlled the region, complained about the invasion to the khedive and the British. Eventually, diplomatic pressure compelled Ismail to abandon the venture, and he dispatched orders for McKillop and Chaille-Long to return with their men to Cairo. By that time, the American colonel, seriously ill with fever, was only too willing to comply.[42]

Of the four elements of the Abyssinian campaign, only that of Rauf Pasha achieved any measure of success. His troops beat off attacks by local tribesmen, and by mid-October had captured Harar. There, Rauf enjoyed the spoils of his conquest and reigned as governor.[43]

The Abyssinian reverses disturbed the khedive and embarrassed Stone. Trying to save face, the chief of staff maintained that a single defeat, or even a string of defeats, did not mean the loss of the war. How many times had the Union army been whipped before achieving ultimate victory? What had to be done was to reorganize, rearm, and develop plans for a new invasion. That was just what Stone set about doing. The khedive's blood was now up, with his reputation as an empire builder at stake. Nevertheless, Ismail disapproved of the plan developed by Stone and favored by Foreign Minister Nubar Pasha for an all-out attack to destroy the Abyssinian army in favor of a punitive expedition that would teach Yohannis a lesson, followed by negotiations on the border problem and other issues.[44]

In selecting a commander for the campaign, Stone favored one of his Civil War veterans, while the Turco-Circassian military leadership promoted one of their own. In the end, the latter group succeeded in convincing the khedive that the leaders of the three unsuccessful prongs of the previous campaign had been foreigners, so he selected a Circassian, Muhammud Ratib Pasha. Disappointed in this, Stone was able to choose an officer to be second in command and chief of staff. He asked Dye, but the colonel refused to serve directly under an Egyptian officer. Stone next approached Loring, who accepted, thinking he would really be in charge. Ten other Americans left with the army as well, including Dye, who became adjutant general, Captain David Essex Porter, the son of Admiral David Dixon Porter, James Dennison, one of the survivors of Arendrup's ill-fated command, and Major Charles Loshe, who had the thankless tasks of trying to maintain the expedition's quartermaster, commissary and transportation departments.[45]

What became known as the Gura Campaign, because of the location of its principal battle, got under way in December 1875. The army landed at Massawa and followed basically the same route into Abyssinia as that taken by Arendrup. Upon reaching the Gura Plain in central Abyssinia, they learned of the presence of a strong enemy force nearby. Instead of heeding the American advice to attack at once, Ratib ordered his men to entrench themselves in strong forts and wait for the Abyssinians to come to them. When this did not happen, on March 7, 1876, a force of about six thousand Egyptians left the safety of their fortifications to meet the Abyssinians. Poor deployment of the battalions on unfavorable ground led to disaster. Unrelenting Abyssinian attacks accounted for over 4,500 Egyptians killed, wounded, or captured. The survivors holed up in a redoubt called Fort Gura, while the Abyssinians halted to loot the bodies that littered the plain and search for food. After beating back another attack, the Egyptian remnant escaped. They joined a brigade that had been guarding the road to Massawa, as well as other reinforcements. The Egyptians now had ample troops to renew the campaign, which is what the Americans wanted. Ratib Pasha and his men, however, had had enough. They pulled back and dug in at Massawa. An order from the sultan at Constantinople to furnish soldiers for a war in the Balkans provided the pretext for an honorable withdrawal back to Cairo. There would be no third attempt against Abyssinia.[46]

Trying to put as good a face on the defeat as possible, the khedive awarded decorations to the heroes of the Battle of Gura, as well as Stone. Stone went along with the charade, but he knew better. Gura had been a debacle, pure and simple. The Americans blamed the Egyptians for cowardice and bad generalship, and the Egyptians blamed the Americans for providing bad advice, but there was enough blame to go around. Ismail knew the truth about Gura as well, and for him the bloom was definitely off the American rose. Stone and his general staff had promoted and planned the campaign and his officers held positions of authority under Ratib. The ex-Yankees and Confederates had done good service in exploring and mapping his empire, but to Ismail the soldiers they had been responsible for training and equipping had proven ineffective in the face of poorly armed tribesmen. The status of the Americans, therefore, became more tenuous. Stone did manage to ingratiate himself again with the khedive and defended the actions of his men. Some of his officers continued their engineering and cartographic work while others, disgusted with Egyptian service, went home. The remainder, having little to do, simply sat around to see what would happen next. They did not have long to wait.[47]

The Abyssinian fiascoes had cost Ismail considerable prestige and money. Prestige he could recoup, but money was another matter. Egypt was deeply in debt, and reports in 1876 by the French and British governments revealed considerable extravagance and mismanagement in the country's finances. As a result, Britain and France pressured the khedive to allow them to control his nation's financial affairs. The following year, again bowing to European pressure, Ismail imposed drastic cuts to public programs and the army and reformed his government. He became a constitutional monarch, and allowed British and French ministers to serve in his cabinet.

The growing European influence in Egypt spelled the end for the American military mission. Neither Britain nor France had any desire to see the Americans continue their service. The downsizing of the army provided the perfect excuse to get rid of them. On June 30, 1878, "Dismissal Day," the last twelve of Stone's officers received orders terminating their contracts. They were given six month's severance pay and traveling expenses back to the United States. By the end of the year, only Stone remained.[48]

The reasons that Stone remained are complex. His name was not on the list of officers slated for dismissal on June 30. Despite his role in the Abyssinian mess, the khedive still liked and trusted Stone. He was only too happy to keep him on as principal military advisor and chief of staff, although there was not much of a staff left to manage. For his part, Stone chose not to resign and leave with his colleagues for financial—he was still in debt and needed the money—and personal reasons. He saw his Egyptian service as a means of washing away the stain of Ball's Bluff. He remained fiercely devoted to Ismail, once telling an associate, "I am perfectly willing to cast my lot with His Highness for good or ill." The khedive recognized and appreciated this so much, that when facing his own dismissal by the sultan at Constantinople (who had the backing of the British and French) on June 25, 1879, he personally asked the general to stay on to assist his successor and eldest son, Tewfik. Ever loyal, Stone, of course, agreed.[49]

Stone easily transferred his loyalty from Ismail to his son. His service, however, differed from that which he had previously provided. There would be no more expansion into Equatorial Africa. Exploring and mapping would be curtailed, and funding was withdrawn from the library, printing office, military academy, and staff school. Education for enlisted personnel also was eliminated. It must have pained Stone to watch as these and other programs he had instituted were dismantled in the name of fiscal responsibility. Nevertheless, he continued to advise Tewfik and traveled abroad as the new khedive's representative. The Stone's home also remained a center for the American community and a place where visitors from the United States could expect a warm welcome. A high point for Stone came when he served as Egypt's delegate and commissioner general to the International Exhibition at Venice in 1881. There he received a decoration as a commander of the Order of the Crown of Italy for his aid to Italian explorers in Africa.[50]

Back in Egypt, Stone's protocol duties also continued for the new khedive. When King David Kalakaua of Hawaii made an unannounced visit as part of his journey around the world in 1881, Stone was among the dignitaries who hastily assembled to greet the monarch at the Cairo railway station. There he was surprised to see an old acquaintance from Civil War days, Hawaiian-born William N. Armstrong, now the island kingdom's attorney general. The two men enjoyed a reunion and over the next several days spent much time together, as Tewfik assigned Stone to be the escort and guide for the Hawaiian royal party. He explained to Armstrong that the khedive had learned through newspaper accounts of the king's travels in the Orient, and, although Egypt and Hawaii did not enjoy formal diplomatic relations, his highness was determined to extend his fellow monarch all of the "courtesies which other nations had shown," including the services of his chief of staff.[51]

Stone arranged a military review for the visitors, as well as the obligatory trip to the pyramids and the sphinx. En route to these monuments the king inquired not about ancient Egypt but about the 1798 Battle of Embabeh, in which Napoleon had decisively defeated the Mamelukes. A startled Stone quickly pointed out the battle site and related how it was fought. After the general's detailed explanation, King Kalakaua remarked that the Mamelukes had been betrayed by poor generalship, a conclusion with which Stone readily agreed. On their return to Cairo, Stone toured the royal party through the old exotic city, with which the general was all too familiar. Armstrong later described the experience:

> We saw what the tourists usually see in the old city; the dreamy, weird life of the Arabian Nights; laden camels picking their way through narrow streets with their heads high in the air; deformed beggars, and almost naked water carriers; maniacal-looking dervishes tossing up their arms; veiled women in shrouds gliding by; and in the narrowest streets, crowded with donkeys and camels, great disturbance and frantic efforts to make way for our carriage.[52]

At the Citadel, Stone took his guests to its imposing alabaster mosque and quadrangle, where Tewfik's great-grandfather, Mohammed Ali, had ordered the massacre of the Mameluke nobles in 1811. After dinner, Stone toured Armstrong through the army headquarters. The Hawaiian quickly grasped the irony of his host's position in Egypt: "Here was an American, of a nation not a hundred years old, teaching a race that had been involved in wars for fifty centuries, how to fight."[53] The general also took the opportunity to explain the unrest among the Egyptian people and military that was growing in opposition to European influence over their country.

Following a visit to the Egyptian Museum to view the "mummies and things," Stone accompanied the king and his party to Alexandria, where the khedive was in residence. The monarchs soon struck up a friendship, perhaps because of what they had in common: "two rulers over nations lying in the way of the Anglo-Saxon march."[54] Through the series of lavish state dinners, balls, and receptions that followed, Stone remained one of King Kalakaua's "constant companions," until the royal party sailed away on a steamer bound for Italy.

The occasional visiting dignitary aside, Stone's time in service to Tewfik would be consumed with the protracted military revolt led by Lieutenant Colonel Ahmed 'Urabi. The son of a village sheik, 'Urabi joined the army as a boy and rose rapidly through the ranks. At age twenty he was a lieutenant colonel. That, however, was as high as the Turco-Circassian military aristocracy would allow a native Egyptian to advance. By the late 1870s, this policy had become increasingly rankling to the Egyptians who saw themselves as just as competent as the Turco-Circassians and foreigners who had led their army to defeat in Abyssinia. In 1881, Tewfik, at the urging of his European advisors, again sought to reduce the size of the army and fired over two thousand officers, most of them native Egyptians. This brought on a rebellion, touched off by the arrest and attempted court-martial of a group of officers, led by 'Urabi, who had submitted a petition to the khedive criticizing the new policy and asking for the dismissal of the war minister. Stone was to preside over the trial, but a group of enraged soldiers broke into the jail and rescued their officers before it could convene. A frightened Tewfik acceded to the demands of the rebellious officers, and for a short while things quieted down.[55]

'Urabi, now the commander of one of the elite guard regiments in Cairo, however, continued to foment trouble. He was the leader of a growing reform movement known as the Nationalist Party. With their rallying cry, "Egypt for the Egyptians," the Nationalists sought to limit foreign influence in their country. They soon constituted a serious threat to the khedive and his European handlers.

In September 1881, 'Urabi and his fellow colonels learned that the new war minister was scheming to stamp out their insubordination. In response they marched their regiments to the Abdin

Palace on the ninth, and there loudly demanded the dismissal of the khedive's entire foreign-dominated ministry, along with an enlargement of the army. When Tewfik appeared before them Stone was at his side. On the advice of Stone and Acting British Consul General Auckland Colvin, the young khedive attempted to woo the ranks away from 'Urabi and the colonels, but the troops would not be swayed. 'Urabi then informed the monarch that his successor could easily be located. The khedive had no other choice but to give in.[56]

After the Abdin showdown, 'Urabi became war minister, and Stone's position became even more tenuous. He acted as much as a statesman as soldier, attempting to steer Tewfik's policies along a very narrow road between the detested, but powerful, European governments on one side, and the wildly popular 'Urabi and his "Egypt for the Egyptians" Nationalist Party on the other. Politics, however, had never been Stone's strong suit.

The American consular mission, following "Dismissal Day" the most influential American body in Egypt, favored 'Urabi. Consul Elbert Farman called him, "the idol of the people," and along with his fellow diplomats, viewed him as a patriot and the best chance for heading off a European occupation of Egypt. On November 11, Stone hosted a meeting in his home between the newly arrived consul from Washington, Simon Wolf, and 'Urabi. Wolf expressed sympathy for the Nationalist cause and urged 'Urabi to be wary of the British and French (which the colonel already clearly understood) and to exercise moderation. 'Urabi promised the Americans wisdom and restraint, but in reality, the time for moderation had come to an end.[57]

For Britain and France, 'Urabi's growing power was intolerable. He had to go, and they let the khedive know it. But there was little Tewfik could do. The publication in January 1882, of a note from the British and French governments threatening intervention in Egypt if 'Urabi was not removed sent events spinning out of control. 'Urabi and the Nationalists, along with the Egyptian people, were enraged, and yet amid this turmoil a singularly peaceful event took place in Cairo. To celebrate Washington's Birthday, Stone and the United States consul general hosted a dinner at the Grand New Hotel. Present were the khedive's cabinet, including 'Urabi, Ferdinand de Lesseps, and other prominent members of the European and American communities. A guest later described the scene:

> After an hour of pleasant conversation, the company adjourned to the large dining-room, which had been decorated with flags, in which those of America and Egypt were everywhere conspicuous. The tables were loaded with flowers. During the whole evening, the band of the Khedive, which were stationed under the windows, played American airs.[58]

Stone began the after-dinner ceremonies with a toast to "the Memory of Washington," which was followed by similar offerings in praise of both the United States and Egypt. The band played "Hail Columbia" and "The Star-Spangled Banner," and the cosmopolitan event concluded, the guests making their way through Cairo's darkened streets, imbued with good feelings.

The optimistic mood did not last long past the Washington's Birthday festivities. On May 25, Britain and France ordered 'Urabi's dismissal as war minister, hoping this would dissipate his influence in the government. A compliant Tewfik agreed, but the army demanded his reinstatement, making the khedive a virtual prisoner in his own palace. Once again, Tewfik had no other alternative but to rescind his order. To the Egyptian populace 'Urabi had won a signal victory for their country and Islam over the foreigners and infidels. Outbreaks of violence against Europeans and Christians began occurring, and Tewfik began to see Britain and France as the only guarantors of his reign.[59]

While internal strife prevailed in Egypt, Britain sent a war fleet to Alexandria. In response, the Egyptian army began strengthening the port's defenses that had been developed years earlier by Loring. For weeks, tensions grew as the two sides eyed one another, the British sweltering in their iron ships, and the Egyptians on the parapets of their forts. On June 11, a riot between Muslims and Christians broke out in Alexandria, resulting in casualties, including foreigners, on both sides. The Europeans claimed that the Egyptian army had been held back for hours before being sent in to restore order. The Nationalists shot back that the European lackey Tewfik had instigated the violence, hoping to discredit 'Urabi and possibly provide a pretext for intervention. At almost the last minute, the sultan at Constantinople, who saw the crisis as possibly resulting in his loss of Egypt, sent a special commissioner to Cairo. All he could do, however, was to convince the khedive to leave the capital for the relative safety of Alexandria, which he did on July 6.

Accompanying the khedive to Alexandria was the ever-loyal Stone, intent on carrying out his "peculiar official responsibilities to the sovereign and Government of Egypt" to the last. On this journey, Stone brought along his thirteen-year-old son, John. Jeannie and the girls remained in Cairo, where they feared for the boy's safety. But Stone insisted that John would be traveling with his father, who was still a man of respect in the Egyptian court. Besides, he assured the family that they would return in a couple of days.[60]

By the time Stone and "Johnny Boy" arrived in Alexandria, however, the situation had gone from bad to worse. It was generally assumed that the British squadron under Admiral Beauchamp Seymour would commence bombardment of the city at any time. Despite John's reassuring telegram to his family on July 7—"ALL WELL"—Stone was worried. The khedive and his entourage had taken up residence in a palace out of range of the British guns, but if fighting broke out, who knew what could happen. It would also be impossible for him to return to Cairo, where anti-foreign agitation could be expected in the event of war. On July 10, Stone made arrangements to place his son aboard the USS *Lancaster*, an American sloop of war that had arrived off Alexandria to offer protection to American and other foreign civilians. He then rushed back to Tewfik's side.[61]

The same day that Stone secured "Johnny Boy's" safety, Admiral Seymour issued an ultimatum: if the Egyptian army did not cease work on its fortifications, he would commence bombardment in twenty-four hours. This alarmed Stone. Although most Europeans and Americans had evacuated Alexandria, there were still many Egyptian civilians in harm's way. Many foreigners, including his own family, also would be stranded in Cairo and in danger of retaliation from enraged Egyptians. He went to local British officials, pleading for an additional twenty-four hours to allow civilians to leave Alexandria and foreigners to get out of Cairo. Their response was to make light of the "heavy gun practice" about to occur. Stone exploded, "If harm shall come to my family, I will kill Sir Beauchamp Seymour!" The admiral threatened to have Stone arrested, but the intercession of Chaille-Long, who was now a provisional consular agent in Alexandria, kept the general out of jail. Stone then had to choose whether or not to warn his family about the upcoming battle. He decided that if word got out, it would likely set off a panic among Europeans in Cairo. In such a melee, he believed Jeannie and the children would stand little chance. He instructed them instead to stay put until the excitement quieted down.[62]

True to his word, Seymour began shelling the Alexandria fortifications at 6:55 A.M. on July 11. His eight ironclads and five gunboats pummeled the outer bastions all morning and for most of the afternoon. The Egyptians, for their part, mounted a stubborn defense, scoring a number of hits on the British ships. By four P.M., however, their guns had been silenced. Shells continued to fall for another hour and a half, when all firing ceased. Most of the forts had been demolished, and hundreds of soldiers lay dead or wounded. In the city, buildings were damaged and fires had broken out. 'Urabi's troops evacuated the next day and rioters took over, looting and attacking foreigners wherever they found them. As Tewfik's official "reporter," Stone made his way to assess the situation. He visited ruined forts and the Egyptian field hospitals to count the dead and attempt to ensure that the wounded, whom he found, "lying on the bare stone floors, covered with blood and dust, gasping for water," received proper care. He also called on what local authority remained to try and prod the prefect of police to restore order. Stone then returned to the khedive to report on conditions in the city and provide whatever counsel he could.[63]

Over the next week, British sailors, soldiers, and marines came ashore to suppress the rioting, address the fires, and offer protection to the khedive and his court. Tewfik was now firmly aligned with the British against his own army. Stone may well have had mixed feelings about this. He definitely did not want to see an expansion of British influence in Egypt, but 'Urabi had committed insubordination bordering on treason. For this, he must be punished. In the meantime, 'Urabi's main force had withdrawn from Alexandria and dug in outside of town. It still constituted a threat. His supporters also held sway in Cairo, making communication between the two cities problematical, and this was of particular concern to Stone since it was almost impossible to keep in touch with his family.

From a military standpoint, Stone also could not obtain accurate reports on 'Urabi's whereabouts and movements. Finally, his officers employed a method he had taught them for getting messages through the Egyptian lines. Writing almost five years later as the "Sometime Chief of the General Staff of H. H. the khedive of Egypt," Stone recalled the situation in mid-July 1882, and how the training he had given his Egyptian subordinates paid off:

> I was with a very few officers and a small body of troops, with the Khedive in Alexandria, while my family were in Cairo, in the midst of the Rebel Government and at their mercy. The trained Egyptian Staff Officers, who had been educated and trained under my orders, remained almost to a man faithful to the Khedive, but they were nearly all caught by the events of the rebellion, within the rebel lines and could not come to us. When it had become evident that Arabi was in open rebellion, and that the only dependence of the Khedive for re-instatement in the possession of the country lay in the British forces which had landed, it was of the greatest importance that we should have reliable information concerning the strength and disposition of Arabi's troops and obtaining of such information was, of course, exceedingly difficult. The few faithful officers and men who succeeded in passing the rebel lines and reported for duty at our headquarters gave us clear and distinct information as to what they had seen, but they generally knew the positions of the troops with whom they had been serving and could give but little reliable information as to the general dispositions and plans of the rebel leaders. Such was the condition of affairs, when one day, a letter (open) written by the senior officer of my staff at Cairo was brought to me by a messenger who had passed through Arabi's Headquarters at Kafr-Dowar, and this letter had been read to Arabi-Pacha, himself. The letter simply stated that he had to inform me that my family were well, that he himself was well, that he was then in the War Department at Cairo and was determined to act the part of a true Egyptian! Arabi-Pacha construed the last phrase to mean that the Colonel was on the side of the rebels and that he was thus giving me bold notice of the fact. He allowed the messenger to go on and deliver the letter to me. When I received and read the letter, the messenger said: "Colonel _____ Bey told me to give this to Stone-

Pacha," and he handed me *a little pocket pin-box.* Remembering the careful instructions I had given to staff officers about the many ways in which dispatches might be secretly carried, I felt assured that the little pocket pin-box contained the real letter, and without opening it went directly to His Highness the Khedive and told him that I had received valuable information from Cairo. His Highness eagerly asked what it was, and I handed him the first open letter, which he read, and then with evident disappointment he said: "Why, General, there is nothing of value here excepting the good news respecting the safety and welfare of your family, which I am happy to see, and it would seem from this letter that even Colonel _____ Bey, whom I so trusted is loose in his loyalty." I replied: "Your Highness, Colonel _____ Bey means by stating to Me that he proposes to act the part of a true Egyptian, that he remains faithful to his duty to Your Highness. The valuable information is here!" and I handed him the little pin-box. The Khedive opened it. It was, apparently, full of pins—and he again looked disappointed. I took the little case and drew out the pins. There was a roll of thin silk paper within. . . . I drew forth the paper and unrolled it. *It contained a full and clear statement of the rebel forces, their positions and the intentions of the rebel leaders!* You can easily conceive how valuable such information was at such a time.[64]

Although the message in question had indicated that all was well with Stone's family in Cairo, they were at the mercy of 'Urabi's supporters. Once Alexandria had been attacked, their position became even more precarious. Almost all of the Europeans and supporters of the khedive had fled the city. Their neighborhood, once the pride of Egyptian society, was deserted. Jeannie, nevertheless, continued to obey her husband's instructions and stayed put. She and the girls did enjoy the protection of a few loyal staff officers and soldiers. As one of their protectors told her, "General Stone is the father of the staff; we will protect you with our lives." Another was even more effusive, proclaiming:

We never had a friend until Stone Pacha came to Egypt. He took us from poverty and wretchedness, and made us what we are, happy, well-fed, well-dressed men, with our families, living in comfort. We swear . . . that no harm shall come to the Pacha's wife and children until we lie dead on your door-step.[65]

But the staff officers and orderlies were few and the rebels many. Trains daily disgorged fanatics shouting, "We have come to teach you Cairenes how to kill Christians." In explaining their predicament to the children, Jeannie minced few words:

> My children, we are in great trouble, but we must look it brave-ly in the face, and try to help each other to bear it. Papa has a good reason, of course, for leaving us here; he may rescue us yet; only we may have to undergo great suffering in the meantime. You know he left me money enough only for a few days' expens-es. That is all gone, and I must use your little store; I shall be forced to exercise great economy, as it will last but a short time. Now, I want you to promise me to be patient, to be cheerful, and always brave. Go on with your studies, keep always busy, and trust to me to save you, if it is possible, when the worst comes. We have fire-arms enough in the house to defend ourselves until we can get help from the staff-officers; and if they fail us, you can be brave and face death like good soldiers. Only prom-ise me never to let an Arab touch you. When it comes to that, remember I expect you to save yourselves by putting a bullet through your heart. Don't leave me to do it.[66]

When word reached Cairo that the khedive had accepted British protection, their situation worsened. Everyone knew that Stone was his most loyal officer, and now that His Highness had allied himself with the invaders, the Stone family, too, would be hated. Their servants could be overheard whispering, "The Bashaw has gone over to the English," and calling them "dogs of Christians." Venturing outside the grounds of their residence also became an adventure for the Stone family, and not a pleasant one. Rebels and ruffians flooded into the city and a walk or carriage ride risked ver-bal insults and physical attack, or even seizure by the police, who had cast their lot with 'Urabi. Every evening Egyptian children gathered in the streets to parade, shouting, "Long live Arabi! God give him victory! Death to the Christians!" Eventually, the Stones became virtual prisoners in their palace home.[67]

For her part, Jeannie remained resolute and determined never "to show the white feather." When confronted by truculent ser-vants who made not-so-veiled threats against her and the girls, she declared:

There never lived an Arab who could frighten me. No, not Arabi and all his troops can do it. Go to your work, you miserable cowards, and the first time you *look* insolent I will have you thrashed. Never dare to threaten me again until you are beyond my reach![68]

She became even more agitated when two of the staff officers, despite their earlier pledges, informed her that they had been ordered to report to 'Urabi's army, leading to the following exchange:

"I am a woman," said she; "but rather than obey an order of Arabi Pacha that would compromise my husband's fidelity to the Khedive, I would let them kill me. You are not faithful soldiers. I cannot understand how you can go. I was not brought up to understand fidelity in this way."

"Madam," said one of them, "they would not accept my resignation; they would shoot me, and how would that help the Khedive or me?"

"It would not help the Khedive," said she sternly, "but it would save your honor."

"His Highness will understand that we were forced to go when Arabi called for us."

"And," said she, "will you dare to face His Highness and give the same excuse that will be given by every traitor in Egypt?"[69]

Despite their mother's pluck, rumors of Stone's death in Alexandria and massacres of foreigners throughout the country kept the family on edge for over a fortnight.

At length, the servants departed, and Jeannie and the children took over the household chores. This was a new experience for the younger girls who had been used to domestics handling everyday duties. Fanny noted in her diary, "I have never worked so hard in my life. Mamma is always on the watch, to see that we are not idle; and even when she reads or plays on the piano for us, we are not allowed to 'hold our hands.' " She concluded with the nonchalance of youth, "How I shall enjoy being lazy by and by, if the Arabs do not kill us before the war is over."[70]

At last, Jeannie received a letter from her husband smuggled through the Egyptian lines. The general and John were safe with the khedive at the Ras-el-Tin palace. He told the family to remain calm, and he would rescue them soon. While the rescue never came, conditions in Cairo did begin to improve. 'Urabi sent Jeannie the general's regular monthly pay of fifty pounds together with a complimentary note. This was, indeed, an ironic twist. 'Urabi claimed to be the head of the Egyptian army, while Stone still occupied the position of its chief of staff. Through some convoluted logic, it was determined that, even though he served as advisor to the khedive who was now a British puppet, he deserved his salary. Things calmed down outside their home as well. The city's prefect of police managed to run the worst of the ruffians out of town, and the girls resumed their daily walks through the deserted neighborhood.[71]

By the end of July, Jeannie had decided it was time to abandon Cairo. She wrote 'Urabi requesting permission to leave and asking for an escort to Ismalia on the Suez Canal. The girls began packing, and five anxious days later, 'Urabi's under minister of war arrived to tell the Stones that the army would provide a special train and escort for them to Ismalia and on to Port Said at the entrance to the canal. Evidently, one of Stone's friends had intimated to 'Urabi that the commander of the American naval squadron was determined to secure the family's release. The colonel, who did not need any more enemies, agreed to let them go, and on August 5, the under minister arranged an elaborate leave-taking for Jeannie and the children from Cairo.[72]

When the family boarded a Russian steamer at Port Said for the overnight voyage to Alexandria Stone was there to greet them. A joyous reunion ensued on the cruise to the ancient port. As the ship arrived off Alexandria Harbor flying the American flag and Stone Pasha's own pennant, the British ironclads and the steam corvette USS *Quinnebaug* fired salutes. Johnny Boy, who had remained in the city in the event his mother and sisters arrived by other means, ventured out aboard a launch. The family was, at last, together. For the interim, they would stay in Alexandria in an undamaged palace belonging to the United States consul.[73]

With his wife and children safe in Alexandria, Stone could return to His Majesty. Although Britain clearly was the power in the land, and his own influence would soon wane, for now Stone remained Tewfik's principal military advisor. He stayed with the khedive as a British army under Lord Garnet Wolseley began its campaign to crush 'Urabi and the Nationalists. After a series of engagements under the brutal Egyptian sun, Wolseley decisively defeated 'Urabi near the Tel-el-Kebir railway station west of Ismalia. The colonel fled to Cairo, where he and his staff offered to surrender to the khedive. Tewfik refused. On September 17, the British entered the capital and made 'Urabi a prisoner to be court-martialed and exiled.

Once the 'Urabi revolt had been suppressed and it was safe to return to Cairo, Stone commanded the special train that transport-ed the khedive back to the capital. He also was likely near Tewfik, if not by his side, for the great review of British troops in Cairo on September 28. Stone must have seen the irony in that Ismail's dream of separation from the Ottoman Empire had at last come to pass, although Egypt was trading subjugation by the sultan for sub-jugation by the queen. Stone did not like the British very much. He never had. In his mind the Americans had come to Egypt to help the country achieve independence and modernization. The British had come to help themselves. He remarked to a visiting American clergyman that the British clerks and office holders arriv-ing in Egypt "swarm upon us like a plague of locusts, and eat out the substance of the land. No wonder that intelligent Egyptians are indignant."[74]

Nevertheless, over the next few weeks he resumed his staff work, trying to put the Egyptian army back together for its new masters. He also began assembling maps and plans for a campaign against Mohammed Ahmed, the Muslim zealot known as the Mahdi, who was stirring things up in the Sudan.[75]

Despite his hard work and loyalty, Stone's influence with and access to the khedive waned rapidly. When British authorities informed him that they would not be undertaking an expedition to suppress the Mahdi, at least not now, he realized that his useful-ness to Tewfik and Egypt was at an end. He knew that once in Egypt, the British would not soon leave. Real independence would remain an illusion. So the family packed up its possessions and bade farewell to the Cairo palace for the last time. Before Stone left, he was given the Order of the Medjidie by the sultan in Constantinople. Tewfik also presented him with the Khedive's Star medal, an award created by the British for service in 1882, which he accepted gracefully. Then, in February 1883, the last American pasha went home.

LIBERTY'S ENGINEER

S TONE returned to the United States as something of a celebrity. He had left almost thirteen years earlier a disgraced army
officer who could not succeed in the mining business. Now he
was the dashing Stone Pasha, and in the minds of many of his countrymen a hero of Egypt. Stories of the British conquest, slaughter
of Colonel William Hicks and his army by Mahdist fanatics, and
Gordon's mission to Khartoum appeared regularly in the national
press, and Stone often found himself asked for his comments and
opinions.

Stone also found that celebrity status does not always pay the
bills, and there were still bills to pay. He had to find steady employment to supplement the income from occasional magazine articles
and speaking engagements. His renown, however, did make it easier to find a job, and he was quickly employed as chief engineer of
the Florida Ship Canal Company. He had seen the benefits of the
Suez Canal, and the concept of a water route for deepwater ships
across north Florida from the Atlantic to the Gulf of Mexico
appealed to him. It had been dreamed of by the Spanish and
actively promoted since the early 1820s. In February 1883, the
state legislature passed a law incorporating the Florida Ship Canal

Company, a venture of mostly northern capitalists. As chief engineer, Stone directed a survey from Jacksonville on the St. John's River, along an almost straight line to Dead Man's Bay on the Gulf. The canal was to be 230 feet wide and 30 feet deep. He estimated the cost of the excavation, equipment, and harbor improvements to be $46,208,462. Unlike the Kanawha Canal, this project would not compete directly with the railroads, but even in an era of big projects, Stone's estimate was quite a sum. In the end, the promoters simply could not raise the capital.[1]

Stone did not remain long enough with the Florida Ship Canal Company to witness its failure. In less than two months, William Tecumseh Sherman and Ulysses S. Grant again intervened. They had been asked by the American Committee of the Franco-American Union to recommend an engineer to manage the construction on Bedloe's Island in New York Harbor of the foundation and pedestal of a colossal monument to liberty then being assembled by sculptor Frederic Auguste Bartholdi in his Paris workshop. They suggested Stone. When the offer to be the project's engineer-in-chief arrived, he accepted without hesitation. Soon he and the family were ensconced in a house in Flushing on Long Island, not quite the palace to which they had become accustomed, but warm and comfortable nonetheless. In the drawing room were a bust and portrait of Khedive Ismail, along with other mementoes of their time in Egypt. His old friends and acquaintances, including Ulysses S. Grant, also threw a welcome back dinner for him at New York City's Park Avenue Hotel, at which the ex-president extolled Stone's achievements in reorganizing the Egyptian army.[2]

The five-hundred-dollar-per-month salary Stone would be receiving was sufficient to keep food on his family's table and a roof over their heads. But the general also became enamored with the idea of the statue itself. It appealed to his patriotism and romantic nature as well as his sense of irony. He was now chief engineer for the world's foremost symbol of liberty, located just a short boat ride away from the island prison of Fort Lafayette, where his own freedom had been taken away two decades earlier.

Even before taking the job, Stone may have been familiar with Bartholdi and his work. A short time before the general arrived in Egypt, Khedive Ismail had been toying with an idea promoted by the Frenchman of a monumental statue to grace the Mediterranean entrance to the Suez Canal. Entitled "Egypt Carrying the Light to Asia," it was to be a sixty-eight-foot-tall rendering of an Egyptian slave woman with an upraised arm holding a beacon that would also serve as a lighthouse. His Highness chose not to proceed with the statue, but in assembling his general staff office in the Citadel, Stone may have stumbled upon some of Bartholdi's sketches or heard about the project in conversations with Ismail or his court.[3]

Stone took an office at 171 Broadway in New York City and began work on the Statue of Liberty on April 2, 1883. His first assistant was ex-Confederate colonel Samuel Lockett, who had served under him in Egypt. Their first task was to establish a camp for about a hundred men on Bedloe's Island at what had been the army's Fort Wood. There, within the walls of the old "star" fort, Stone had to design and lay the concrete foundation upon which the pedestal and statue would rest. His crews, composed largely of Italian immigrants, dug a huge pit, fifteen feet deep by ninety-one feet square in the middle of the parade ground. Into it, in October 1883, they began the largest single block concrete pour in history, eventually amounting to some twenty-seven thousand tons. In the meantime, Stone also had to fabricate a method of transporting the vast amounts of cement needed for the foundation, and wood for the above ground forms to the site. He built a trestle far enough out from the island into the harbor to allow boats ferrying materials to anchor. A small railroad would carry the cement and lumber from the boats to the fort's wall. Lumber and other goods could then be brought in through the fort's sally port, but the cement was another matter, so he devised a method of hoisting it over the wall and running it down to the pit. The pour lasted until the winter, when the job was suspended to allow the slab to settle. It began again the following year, reaching ground level in March. Work then started on the upper portion, or base, on which the statue's pedestal would rest. This was not completed until summer. Always a stickler for detail, Stone carefully inspected the pour to ensure the quality of the concrete and proper compression and curing. This portion of the statue project alone cost about ninety-four thousand dollars, so he did not want any mistakes. Not everything went smoothly, however. The general more than once quarreled with his cement contractor, and at one point the two almost came to blows on the parapet.[4]

About this time, Stone also engaged in a bit of journalistic jousting over events in Egypt. In June 1884, his daughter Fanny published in *The Century* magazine portions of the diary she had kept in
Cairo during the 'Urabi revolt. In his introduction to the article,
Stone strongly criticized Sir Seymour Beauchamp for bombarding
Alexandria without giving adequate warning for the evacuation of
Europeans from the city as well as other parts of the country: "This
barbarous disregard on the part of the British of the lives of citizens
of all nationalities caused me, as well as thousands of others, fearful anxiety, and caused the horrible death of scores of Europeans—
French, Germans, Austrians, and Italians."[5] The general's statements prompted a response in the magazine from Lieutenant
Commander Caspar F. Goodrich of the United States Navy, who
had been executive officer of the USS *Lancaster* off Alexandria. He
defended the British actions and attacked Stone for abandoning his
wife and daughters in Cairo during the emergency. Stone shot back
that the British had been responsible for the deaths of hundreds of
Egyptian civilians, in addition to the Europeans, and vigorously
denied that he had placed his family in danger. He still believed
that they were safer in their Cairo home than trying to escape
aboard a train crowded with panic-stricken Europeans. He also
refuted Goodrich's assertion that the ships of foreign nations, then
off Alexandria Harbor, did little to secure the safety of their nationals in Egypt. In his claims Stone was supported by a published letter from the commander of the warship USS *Galena*, which had
been there to evacuate Americans, but had departed on the day of
the bombardment following the arrival of the *Lancaster*.[6]

While Stone traded literary broadsides with Goodrich, his workmen were completing the foundation pour. At the same time,
noted architect Richard Morris Hunt was finishing plans for the
project's next phase, the pedestal. A founding member of the
American Institute of Architects, and known as the "dean" of the
profession, Hunt seemed the ideal candidate to receive the
pedestal commission from the American Committee. He had
designed the Tribune Tower, which, when completed in 1875, was
the tallest building in New York City. He also had been the architect for the homes of a number of members of the American
Committee. Another plus, he knew Bartholdi.

After a number of unsuccessful attempts, by August 1884 Hunt
had developed pedestal plans that would complement, but not

overwhelm, Bartholdi's statue. His submission came just in time, because Stone and his workmen were putting the finishing touches on the pedestal base. Also, on July 4, 1884, in Paris the French presented the completed statue, "Liberty Enlightening the World," with its interior support structure designed by Gustave Eiffel, to representatives of the United States. But across the Atlantic there was no pedestal on which to place the monument. Money still had to be raised to build that part of the project, so the Statue of Liberty would stand, towering over Bartholdi's workshop, for months before finally being disassembled for shipment to America.[7]

Although total funding to complete the construction of Liberty's pedestal would not be secured for some time, enough money was scrounged for Stone to arrange a cornerstone laying on August 5. He served as the master of ceremonies for the event, which would be conducted by members of the Masonic Fraternity. For the occasion Stone wore a black cutaway coat festooned with decorations he had received during his stint in Egypt. A driving rainstorm, however, almost washed out the affair, as Stone, the Freemasons, the band, and about five hundred invited guests got soaked. Nevertheless, the six-foot, ten-inch by three-foot, eight-inch by two-foot, six-inch granite block cornerstone for "Liberty's place of rest" was able to be situated with due solemnity on the northeast section of the base. Everyone then dashed for cover. The bad weather would serve as an omen for what would be a most trying year to come.[8]

At the soggy cornerstone ceremony, Stone remarked to a newspaperman that if money for the pedestal was not forthcoming, he would have no choice but to stop work. That would occur sooner than anyone thought. Since the inception of the idea for the Statue of Liberty, the American Committee had been attempting to raise funds for the foundation, base, and pedestal. It found the going tough. America's moneyed elites did not rush to contribute, and Congress refused to authorize any federal support. Outside of New York City, the project failed to stir much interest. San Francisco and Philadelphia politicians thought it a worthy endeavor, but they wanted the statue for their own cities and saw no reason to help New York. Even Governor Grover Cleveland vetoed assistance from New York State. By mid-1884, the committee had raised just enough in its treasury to pay Hunt's commission (which he returned) and Stone's salary (which he did not return), and complete the foundation and base for the pedestal. Over the next year, the committee struggled to bring in contributions. Stone, in the meantime, continued construction of the pedestal on a pay as you go basis, starting when money became available and slowing or suspending work when it looked like it might run out.[9]

Stone had estimated that following the cornerstone laying he had funds sufficient to lay the first twenty feet of blocks of the finest New York granite for the eighty-nine-foot pedestal. Sporadic contributions enabled him to continue off and on into 1885. On March 10 of that year, with only about two thousand dollars remaining in the treasury, however, the American Committee ordered all work stopped.[10]

By the time the last mason packed away his tools, the statue had become somewhat of an embarrassment for the American Committee, the city of New York, and the United States. It looked as though "Liberty Enlightening the World" might remain across the Atlantic indefinitely. For Stone, however, the project had become as much a crusade as a job. He turned into a tireless promoter of the monument. He wrote articles and gave lectures on the ever-popular topics of Egypt and "Chinese" Gordon, all to raise funds. He also claimed he would work, at times, without pay as a donation to the cause. But Stone's efforts were just a drop in the bucket. So were the fancy benefit balls, exhibits, carnivals, and concerts, the sales of stereograph slides of the completed work in Paris and of the forearm, hand, and torch on display in New York's Madison Square, and six- and twelve-inch-high models of the statue standing proudly on its pedestal. The project needed an angel.[11]

That angel came in the form of an immigrant from Hungary, Joseph Pulitzer, publisher of the New York *World* and Saint Louis *Post-Dispatch* newspapers. When the Committee announced that work on the pedestal had stopped, he initiated a fund-raising campaign in his papers. He employed the symbolic image of Uncle Sam holding out his hat for donations, and it worked. Donations flooded in, mostly from New York City and surrounding communities, but a fair number from more distant cities and towns as well. Some were as little as a few cents, but every contribution was duly noted on the pages of the *World*. In two months, $92,000 had come in, and Stone resumed work on the pedestal. By August 5, 1885, a year after the laying of the cornerstone, the remaining $100,000 needed to complete that phase of the project had been raised. Pulitzer's rescue had come in the nick of time. On May 21, the French ship *Isere* steamed into New York Harbor carrying the copper sheets and iron framework of the disassembled statue in over two hundred packing crates. The event captured the public's imagination, and Stone arranged a parade in honor of the arrival in New York City as well as a welcoming ceremony for dignitaries on Bedloe's Island.[12]

Bartholdi had envisioned and sculpted the statue, Eiffel had configured its internal support, Hunt had designed the pedestal, and Pulitzer had raised the money for its completion. It was now up to Stone to put all of these parts together. This responsibility would soon put the general under a critical microscope, but for the time being he could take pleasure in knowing that the project would be going ahead, and revel in the pun offered by Pullman Company vice president Horace Porter at a farewell dinner for the crew of the *Isere*: "We long ago prepared the stones for the pedestal, and we first secured the services of the most useful, most precious stone of all—the Pasha from Egypt."[13]

Over the summer of 1885, while the Statue of Liberty remained in its crates on Bedloe's Island, work on the pedestal dragged on. The slow progress began to cause grumblings about the competence of the project's chief engineer. Stone was criticized by a freelance journalist named Edward Rudolf Garczynski. He claimed that Stone denigrated Bartholdi's effort, quoting him as saying, "He [Bartholdi] has left considerable work of American engineering ingenuity." In particular, Garczynski maintained the general had complained that the French blundered by not addressing the problem of attaching the copper skin to the iron frame, leaving the fault up to him to set right. Although Stone adamantly denied he had ever disparaged Bartholdi or Eiffel, doubt had been raised about his own honesty and ability.[14]

As the pedestal fund-raising campaign neared its goal, public criticism of Stone picked up. The New York *Evening Post* published a front-page story citing an unnamed source who accused the general of padding his staff with unnecessary inspectors and draftsmen. Once again, Stone, backed by Pulitzer's *World*, denied the facts of the story, and promised a full accounting of his staffing and expenditures. In subsequent articles, he blamed his disgruntled former cement contractor for the allegation. This only added fuel to the fire of Stone's critics who now charged him with overspending and incompetence when it came to the cement construction. General G. A. Gilmore of the United States Army Corps of Engineers also chimed in with his opinion that Stone had spent far too much on the foundation and base.[15]

With so much criticism swirling about its chief engineer, a nervous American Committee decided to cut Stone loose. It announced that after January 1, 1886, with the pedestal expected to be complete, French engineers would finish the project. Stone responded with an offer to volunteer his services, but attacks continued on the quality of his work, along with more serious allegations that he had carelessly left the crates containing the statue out in the open and subject to damage from rain and salt air corrosion. Despite his impending dismissal, Stone kept pushing the pedestal toward completion over the fall of 1885. When Bartholdi visited Bedloe's Island in November, he gave Stone an outward vote of confidence. Nevertheless, the Frenchman was disappointed to learn from the chief engineer that the pedestal would not be completed and the statue in place in time for a July 4, 1886, dedication. Rather, he now hoped for September.[16]

In the end, the American Committee reversed its decision and kept Stone on past January 1. There was still construction to be done on the pedestal, and he convinced the organization's leadership that his work and judgement had been sound, and that he was the right man to finish the project. On April 22, Stone once again hosted a special event on Bedloe's Island to mark the completion of the pedestal. Before the final piece of granite was set in place, he announced that he wished to have it encased in silver. The dignitaries present thereupon threw nickels, dimes, quarters, half-dollars, and dollar coins into the mortar before the workmen lowered the block. Stone remained in charge after the pedestal's dedication as well, and began figuring out the best methods for attaching the copper sheets to the frame and the frame to the pedestal. He also asked for a salary of $12,500 to see these, the trickiest and most critical parts of the construction, to a successful conclusion.[17]

Stone at last developed the system for anchoring the frame to the pedestal using steel beams and tension bars sunk into the structure. A method of attaching the sheets to the frame, avoiding any galvanic issues, also was devised. Assembling the statue itself, however, was like trying to put together a giant three-dimensional jigsaw puzzle. The framing beams and copper facings had all been numbered in accordance with Bartholdi's and Eiffel's instructions before being shipped. When the crates were opened, however, the workmen found the numbering system to be incorrect. Having to pick through the beams and sheets, identify them, and then match them to the diagrams provided by the French, slowed the process considerably. If Stone had, indeed, believed Bartholdi to have been sloppy in his planning, he may have felt vindicated. Also retarding work on the statue was a problem everyone thought had been solved: money. Pulitzer had, indeed, raised enough cash for the pedestal, but now the committee found its coffers were running dangerously low for actually putting the statue together. At its suggestion, Stone sent out a circular to the "200 wealthiest men and women" of New York City and Brooklyn, asking them to donate the fifteen thousand dollars needed to complete the project. The effort raised three hundred. Finally, on July 10, Congress passed an appropriation sufficient to finish the statue and hold an inaugural ceremony.[18]

With funding assured, much assembly work remained to be done. It became clear that a September completion date would be impossible. The inaugural, therefore, was pushed back to October 28, 1886. While his workmen, under the guidance of landscape architect Frederick Law Olmsted, tidied up the construction site, Stone oversaw the last phase of the project. He contracted with a lighting company for the lamps and machinery needed to allow Liberty's torch to function as a lighthouse. This had been required by the federal government. In addition, he positioned incandescent lamps in the crown to give the appearance of jewels and placed strong lights at the base to illuminate the folds of Liberty's drape. Stone, however, received quite a bit of criticism for his lighting plan. He replied that his principal motivation in lighting Liberty had been aesthetic and not navigational:

> The arrangement for lighting the statue is the result of long and careful study, and I think it is the best that can be devised for this purpose. It is intended to light up the statue so as to make the entire work visible at night. The lights will not be so

arranged that they will dazzle the eyes of persons looking at the statue. If they were arranged so as to throw a powerful illumination down the Bay, the statue itself would be in total darkness. It would not be visible. That is not the design.[19]

In what became an example of act first and ask permission later, Stone realized that the government could take weeks to review his plan, and began work on the lighting without getting prior approval. Nevertheless, it was not ready in time for the inauguration, and the statue never did function properly as a lighthouse.[20]

During the congressional debate over the appropriation for the statue, Stone had insisted that it include money for an elaborate platform for the inaugural ceremony and $2,500 for refreshments. Although the lawmakers declined to fund the general's requests, October 28 would see an appropriately impressive event. President Grover Cleveland appointed Civil War veteran and commander of the army's Department of the Atlantic, General John Schofield, to direct the inaugural, and he, in turn, named Stone to be the grand marshal of the day's "military, civil and naval parade" in New York City. The parade would march through Manhattan in the morning, with the actual unveiling of the statue not occurring until the afternoon.[21]

In planning the great parade, Stone reverted to his old army habits. His system of organization followed along military lines, with participants arranged in ranks and divisions. Instructions and updates appeared in the form of general orders. Stone issued a circular in which he invited all military and civilian associations, clubs, and other groups, as well as representatives from the states and territories, Canada, British Columbia, Mexico, and the West Indies to join in the review. He expected a moderate response. So great was the reaction from the public, however, that to get any work done he had to lock himself in his office to get away from the throng of registrants at his door clamoring for permission to join in the festivities.[22]

The weather in New York had been rainy up until the inaugural, dredging up memories for Stone of the soggy cornerstone ceremony. October 28, however, dawned cold, windy, overcast, but—for the time being—dry. As the hordes of spectators gathered along the parade route a drizzle did begin to fall, but nothing could spoil the general's big day. Once again, he donned his medals and decorations. Stone appeared at the starting point mounted on a white charger. He was bareheaded, his white hair, beard, and moustache matching his stallion. Behind him some thirty thousand marchers—bands, military, police and fire units, Civil and Mexican War veterans' organizations, fraternal societies, schoolchildren, and patriotic, civic, and church groups, waited anxiously to get underway. When President Cleveland arrived at the reviewing stand at Fifth Avenue and Waverly Place at a little after ten A.M. the parade finally began. Stone initially led the march, but as it passed the president, he peeled off and positioned himself to the right of the reviewing stand for the duration of the two hour procession. After the last contingent had gone by, Schofield turned to Stone and said, "This is the best parade I have ever seen in New York."[23]

Stone and Schofield, along with President Cleveland and a host of dignitaries, then made their way to Bedloe's Island, where the inaugural ceremony and unveiling would take place. By now the rain had picked up. Cannon salutes and whistles from the hundreds of ships and boats that crowded the harbor around Bedloe's created a cacophony that made it difficult to hear what was going on. Finally, using his command voice, Schofield shouted for the event to begin. After the invocation, Senator William Evarts of New York, chairman of the American Committee, launched into his speech presenting the statue to the United States. The plan was for the veil that had been covering Liberty's face to be dropped at the conclusion of his address. A pause midway through, however, was mistaken for the ending, and down went the drape. Another, even noisier, round of cannon fire, whistles, and cheers lasted for a half hour. Evarts gave up and handed the title to the statue to Cleveland. Obviously upset, he turned to Stone sitting nearby, who could only shrug his shoulders. Once order had been restored, Cleveland, Bartholdi, and the French delegation offered their remarks, followed by the closing benediction. The dignitaries and spectators then dispersed from the island to dinners and receptions throughout Manhattan. The massive fireworks display scheduled for that evening had to be postponed due to the rain.[24]

The next morning, Stone slept in. He did not arrive at his office until 11:30 A.M., telling reporters, "I thought I would take things easy." But there were still matters pertaining to the project that needed to be wound up, not the least of which was the lighting, which would not be completed until November. Even though Schofield had praised the previous day's parade, Stone still could not escape controversy. It seemed that the mayor of New York City had not been reserved a place on the reviewing stand, and the governor and representatives of the French fleet were not included as part of the Bedloe's Island program. Responsibility for these and other miscues was immediately placed on Stone. He responded that he had been in charge only of the parade itself, not the reviewing stand, and certainly not the event on Bedloe's. Nevertheless, when questioned by the press about their own shortcomings in the Statue of Liberty ceremonies, other members of the American Committee continued to maintain that Stone was, indeed, to blame for what went wrong. By now he must have been used to being the scapegoat.[25]

Once the squabbles over the Statue of Liberty subsided, Stone and his family entered into a period of quiet retirement at their Flushing home, although he still maintained an office in lower Manhattan overlooking the Battery. On New Year's Day 1887, he re-emerged from relative obscurity to answer a warning from Major David Porter Heap of the Corps of Engineers that the upraised arm of the statue was in danger of snapping off. Stone was rightly proud of the monument and was quick to refute any "misgivings about the goddess's strength and powers of endurance." He told the *New York Times*:

> Major Heap is stated to think the torch arm will break in the not distant future. There is no statement as to whether he thinks the copper or the iron framework will fail. I do not think either will fail. That the ascent through the arm is uncomfortable is hardly to be wondered at; that it is dangerous for a man of ordinary strength and activity I do not think.[26]

Stone was right. The arm did not fail. It suffered only minimal damage from a massive explosion of a munitions depot at Jersey City in 1916 that shook all of New York Harbor. Tourists continued to climb to the torch until 1917, when the number of visitors wishing to make ascent up the ladder became overwhelming and authorities imposed a ban.[27]

Despite the criticisms that had been leveled against him, Stone felt his work on the project had, at last, redeemed his reputation in the United States. He also must have taken pride in John, who attended the Columbia Grammar and Preparatory School in New York City and then enrolled in the Columbia University School of Mines to study engineering. In early 1887, the Stones began preparing for a trip to New Orleans to visit Jeannie's family. The Louisiana Stones must have thought highly of their Yankee in-law, since Jeannie's brother, John Horace Stone, named his first child Charles Pomeroy. While Jeannie was looking forward to visiting with her relatives, it is likely that the general was going south to look for work. As usual, the family again was in need of money. John Horace was a prominent attorney and judge who could have been able to help his brother-in-law find suitable employment. A few days before leaving New York, however, Stone fell ill at his office and went home with a cold, which quickly developed into pneumonia. An inveterate smoker, his lungs may have been too compromised to fight off the infection. After a week's illness, he passed away at home on January 24, 1887, at 1:30 in the afternoon with Jeannie and the children at his bedside. Charles Pomeroy Stone was sixty-two.[28]

Stone's funeral was held three days later at Saint Leo's Catholic Church on West Twenty-Eighth Street in New York City. Many Mexican and Civil War colleagues attended. Pallbearers included William Tecumseh Sherman, Fitz-John Porter, Charles Devens (who had been Stone's subordinate at Ball's Bluff), and General Schofield. After the service, his casket was placed aboard a special railroad car for transportation to West Point for burial. General and Mrs. Sherman were among those accompanying the family to the gravesite. At West Point a company of army engineers unloaded the casket, placed it on a gun carriage, and escorted it to the United States Military Academy Cemetery. There Stone was laid to rest with full military honors. A marker in the shape of a shield denotes his grave. It bears his Egyptian rank, lieutenant general, along with the phrase, "A Soldier Without Fear, and Without Reproach."[29]

EPILOGUE

W HILE Stone may have regained his good name in Egypt and at the Statue of Liberty, he did not recoup his fortune. Indeed, when his estate was settled, after paying off debts, there remained only a hundred dollars in cash. This left Jeannie, who still had Fanny and Todas at home, in strained financial circumstances. They quickly headed to Louisiana and the support of her family. Coming to her aid was another soldier of misfortune, Stone's West Point classmate Fitz-John Porter. He ended his eulogy for the general, written for the organization of United States Military Academy graduates, by urging his colleagues: "Let us all unite in soliciting that his family shall not suffer because he spent his substance and imperiled his life in the service of his country." Porter also vouched for Jeannie when she applied for a Mexican War widow's pension in October 1887. In June of the following year, she did receive a monthly pension of fifty dollars, the standard amount for the widow of a regular army colonel, through a special act of Congress.[1]

Eventually, Jeannie moved to San Diego, California, to be near Todas and her husband. She died there in 1924. After attending Columbia and graduating from Johns Hopkins, John Stone became an internationally renowned mathematician, electrical engineer, and inventor. Over the course of his career, he obtained a number of patents in the area of wireless communication. Married and divorced, he died in San Diego in 1943. Todas remained in San Diego until her death in 1947. All three are buried in the city's

Mount Hope Cemetery. Jeannie's marker reads simply, "Wife of Lieutenant General Charles Pomeroy Stone."

The later activities and whereabouts of Hettie and Fanny remain unknown.

NOTES

CHAPTER ONE: HILLS OF NEW ENGLAND TO THE WEST POINT PLAIN

1. "Biographical Notice of the Late Dr. Alpheus Stone," *Boston Medical and Surgical Journal* 45 (October 15, 1851): 217–219.

2. Fitz-John Porter, "Charles Pomeroy Stone," *Eighteenth Annual Reunion of the Association of Graduates of the United States Military Academy at West Point, New York, June 9, 1887* (East Saginaw, MI: Evening News Printing and Binding House, 1887), 40.

3. Robert E. Temple to Joel Poinsett, Albany, New York, January 8, 1841; Luther B. Lincoln to Joel Poinsett, Deerfield, Massachusetts, July 1, 1840, United States Military Academy Appointment File, RG 94, Records of the Adjutant General's Office, National Archives.

4. John C. Waugh, *The Class of 1846, From West Point to Appomattox: Stonewall Jackson, George McClellan and their Brothers* (New York: Ballantine Books, 1994), 18–19, 24–25.

5. Ibid., 30–31, 37–39; Donald C. Pfanz, *Richard S. Ewell: A Soldier's Life* (Chapel Hill: University of North Carolina Press, 1998), 18.

6. Lloyd Lewis, *Captain Sam Grant* (Boston: Little, Brown, and Company, 1950), 80–81, 87–88, 91; *Official Register of the Officers and Cadets of the US Military Academy, West Point, New York, June, 1842*, 13–15.

7. Waugh, *Class of 1846*, 9, 26, 29; Lewis, *Grant*, 62–64; Pfanz, *Ewell*, 19–20.

8. *Official Register of Officers and Cadets, 1842*, 2, 13; Waugh, *Class of 1846*, 40.

9. *Official Register of the Officers and Cadets of the US Military Academy, West Point, New York, June, 1843*, 2, 10, 17; *Official Register of the Officers and Cadets of the US Military Academy, West Point, New York, June, 1845*, 2, 3, 7, United States Military Academy Library; Waugh, *Class of 1846*, 39–40, 66; Pfanz, *Ewell*, 25–26.

10. Porter, "Stone," 40.

11. *Official Register of the Officers and Cadets, 1845*, 7; Lewis, *Grant*, 76; Richard Bruce Winders, *Mr. Polk's Army: The American Military Experience in the Mexican War* (College Station: Texas A&M University Press, 1997), 21; Ron Field, *Mexican-American War* (London: Brassey's, 1997), 59–60.

12. Waugh, *Class of 1846*, 74–75; Porter, "Stone," 41; George W. Cullum, *Biographical Register of the Officers and Graduates of the US Military Academy at West Point, NY, Vol. II, 1841–1847* (New York: D. Van Nostrand, 1868), 117; Returns from US Military Posts, 1800–1916, West Point, NY, July 1845, September 1845,

January 1846, RG 94, Records of the Adjutant General's Office, National Archives.

Chapter Two: To Mexico with Winfield Scott

1. John Akelyn, *Biographical Dictionary of the Confederacy* (Westport, CT: Greenwood Press, 1977), 241–242; Trevor Dupuy, Curt Johnson and David L. Bongard, *The Harper Encyclopedia of Military Biography* (New York: HarperCollins, 1992), 354. More information on Benjamin Huger, particularly on his career during the Civil War, can be found in Jeffrey Rhoades, *Scapegoat General: The Story of General Benjamin Huger, CSA* (Hamden, CT: Archon Books, 1985).
2. K. Jack Bauer, *The Mexican War, 1846–1848* (Lincoln: University of Nebraska Press, 1992), 246–247.
3. Josiah Gorgas, "Journal of a Campaign in Mexico," manuscript in the Josiah and Amelia Gorgas Family Papers, W. S. Hoole Special Collections Library, University of Alabama, Birmingham, 1; Frank E. Vandiver, "The Mexican War Experience of Josiah Gorgas," *Journal of Southern History* 13 (August 1947): 376; Winders, *Mr. Polk's Army*, 89–90.
4. Gorgas, "Journal," 1–4.
5. Ibid., 5, 7.
6. Ibid., 8–9; Benjamin Huger, "War Journal of an Ordnance Chief: Being the Mexican War Diary of Captain Benjamin Huger, Chief of Ordnance, Army of Invasion," ed. Jeffrey L. Rhoades, typescript copy in the National Museum of American History, Washington, DC, 3.
7. Ibid., 10; *General Scott and His Staff; Comprising Memoirs of Generals Scott, Twiggs, Smith, Lane, Cadwalader, Patterson and Pierce; Colonels Childs, Riley, Harney, and Butler, and Other Distinguished Officers Attached to General Scott's Army; Together with Notices of General Kearny, Colonel Doniphan, Colonel Fremont, and Other Officers Distinguished in the Conquest of California and New Mexico* (Philadelphia: Gregg, Elliot & Co., 1848), 24; Cullum, *Biographical Register* 2, 117; Paul C. Clark and Edward M. Mosley, "D-Day, Veracruz, 1847: A Grand Design," *Joint Force Quarterly* (Winter 1995–96): 110.
8. Gorgas, "Journal," 11–12; Huger, "War Journal," 12–13; John S. Jenkins, *History of the War Between the United States and Mexico from the Commencement of Hostilities to the Ratification of the Treaty of Peace* (Auburn, NY: Derby and Miller, 1851), 246, 250–252, 256; Edward D. Mansfield, *The Life and Services of General Winfield Scott, Including the Siege of Vera Cruz, the Battle of Cerro Gordo, and the Battles in the Valley of Mexico, to the Conclusion of Peace and His Return to the United States* (New York: A. S. Barnes & Co., 1852), 372–373; Timothy D. Johnson, *A Gallant Little Army: The Mexico City Campaign* (Lawrence: University Press of Kansas, 2007), 23, 25–28.
9. Gorgas, "Journal," 12–13.
10. John Jacob Oswandel, *Notes on the Mexican War, 1846–48* (Philadelphia: n.p., 1885), 80, 84; N. C. Brooks, *A Complete History of the Mexican War: Its Causes and Consequences: Comprising an Account of the Various Military and Naval Operations from its Commencement to the Treaty of Peace* (Philadelphia: Gregg, Elliot & Co., 1851), 300.
11. Huger, "War Journal," 14. After her service as a military transport, the *Tahmaroo* carried passengers between New York and San Francisco during the California Gold Rush and was a whaler in the Pacific.

12. Huger, "War Journal," 14, 16; Winfield Scott to Wm. L. Marcy, Vera Cruz, March 23, 1847, in Brooks, *Complete History*, 302–306.
13. Raphael Semmes, *Service Afloat and Ashore During the Mexican War* (Cincinnati: Wm. H. Moore & Co., Publishers, 1851), 138.
14. Mansfield, *Life and Services of Winfield Scott*, 374.
15. Col. Ethan A. Hitchcock to Elizabeth Nicholls, Vera Cruz, March 27, 1847, in Winston Smith and Charles Judah, eds., *Chronicles of the Gringos: The US Army in the Mexican War, 1846–1848, Accounts of Eyewitnesses and Combatants* (Albuquerque: University of New Mexico Press, 1968), 194–195.
16. Ramon Alcaraz, Alejo Barreiro, Manuel Payno, Guillermo Prieto, Ignacio Ramirez, et al., *Apuntes para la historia de la Guerra entre Mexico y los Estados Unidos* (Mexico, 1848), in Krystyna Libura, Luis Gerardo Morales Moreno, and Jesus Velasco Marquez, *Echoes of the Mexican-American War* (Toronto: Groundwood Books, 2004), 112–113.
17. Mansfield, *Life and Services of Winfield Scott*, 375.
18. Winfield Scott, *Memoirs of Lieut. General Scott, LLD* (New York: Sheldon & Company, Publishers, 1864), 429; Huger, "War Journal," 18.
19. Peter V. Hagner to Mary M. Hagner, "Camp before Vera Cruz," March 10, 1847, in Smith and Judah, eds., *Chronicles of the Gringos*, 196–197.
20. Semmes, *Service Afloat and Ashore*, 164, 167; Huger, "War Journal," 21–22; Gorgas did not accompany the invading army, having been ordered to remain in Vera Cruz and establish an ordnance depot there.
21. Gorgas, "Journal," 17.
22. Huger, "War Journal," 23.
23. Robert Anderson, *An Artillery Officer in the Mexican War, 1846–1847* (New York: G. P. Putnam's Sons, 1911), 153.
24. Cadmus Wilcox, *History of the Mexican War* (Washington, DC: Church News Publishing Company, 1892), 308–309.
25. Ibid., 309–310; Bauer, *Mexican War*, 271; R. S. Ripley, *The War With Mexico*, vol. 2 (New York: Harper & Brothers, 1849), 109–110; While Cullum's *Biographical Register* does not list Amazoque as one of the engagements in which Stone participated, his biography in the "Original Members" section of the Aztec Club of 1847 website does state that he was present in that action: http://www.walika.com/aztec/bios/stone.htm.
26. Carlos Maria de Bustamante, *El Nuevo Bernal Diaz del Castillo, o sea historia de la invasion de los angloamericanos en Mexico* (Mexico: Conaculta, 1990), in Libura, Moreno, and Marquez, *Echoes*, 123; David A. Clary, *Eagles and Empire: The United States, Mexico, and the Struggle for a Continent* (New York: Bantam Books, 2009), 333.
27. Huger, "War Journal," 44–46, 55; Robert Royal Miller, ed., *The Mexican War Journal and Letters of Ralph W. Kirkham* (College Station: Texas A & M University Press, 1991), 35.
28. Simon Bolivar Buckner to his cousin, November 17, 1847, in Johnson, *Gallant Little Army*, 152.
29. Anderson, *Artillery Officer*, 286.
30. Daniel Harvey Hill Journal in Smith and Judah, eds., *Chronicles of the Gringos*, 239.
31. Huger, "War Journal," 69–70; Cullum, *Biographical Register* 2, 117.

32. Huger, "War Journal," 78–80.

33. Charles J. Patterson, *The Military Heroes of the War with Mexico: With a Narrative of the War* (Philadelphia: William A. Leary & Company, 1850), 112–115; Clary, *Eagles and Empire*, 363; Bauer, *Mexican War*, 308.

34. John Frost, *Pictorial History of Mexico and the Mexican War: Comprising an Account of the Ancient Aztec Empire, the Conquest by Cortes, Mexico Under the Spaniards, the Mexican Revolution, the Republic, the Texan War, and the Recent War with the United States* (Philadelphia: Charles Desilver, 1852), 554; Jenkins, *History of the War*, 389.

35. Huger, "War Journal," 86–87.

36. Cullum, *Biographical Register* 2, 117; William Croffut, ed., *Fifty Years in Camp and Field: Diary of Major General Ethan Allen Hitchcock, USA* (New York: G. Putnam's Sons, 1909), 299; Huger, "War Journal," 88.

37. Huger, "War Journal," 89–90; Scott, *Memoirs*, vol. 2, 511; R. S. Ripley, "Chapultepec and the Garitas of Mexico," *Southern Quarterly Review* (January 1853): 17.

38. Alcaraz et al., *Apuntes para la historia de la guerra*, in Libura, Moreno, and Marquez, *Echoes*, 142; Croffut, *Fifty Years*, 301; Winfield Scott to William L. Marcy, Mexico City, September 18, 1847, in Stephen L. Butler, ed., *A Documentary History of the Mexican War* (Richardson, TX: Descendants of Mexican War Veterans, 1995), 231–237; Scott, *Memoirs*, vol. 2, 519; Cullum, *Biographical Register* 2, 117.

39. Huger, "War Journal," 155; Johnson, *Gallant Little Army*, 236–237.

40. Huger, "War Journal," 91–92, 157–159, 162.

41. Winders, *Mr. Polk's Army*, 122, 135, 238. The Aztec Club of 1847 is still active. Its website, www.aztecclub.com, contains a history of the founding of the organization.

42. James M. McCaffrey, ed., *Surrounded by Dangers of All Kinds: The Mexican War Letters of Lieutenant Theodore Laidley* (Denton: University of North Texas Press, 1997), 150; Huger, "War Journal," 135.

43. Huger, "War Journal," 137–138; Miller, ed., *Mexican War Journals and Letters*, 103–108; Ulysses [Grant] to Julia [Dent], Tacubaya, May 7, 1848, in John Y. Simon, ed., *The Papers of Ulysses S. Grant, Volume I, 1837–1861* (Carbondale: Southern Illinois University Press, 1967), 155–156; Grant, *Memoirs*, vol. 1, 180–181; "A Visit to Popocaptetl," *Putnam's Monthly Magazine* 1 (April 1853): 408–416.

44. Huger, "War Journal," 188; Winders, *Mr. Polk's Army*, 145. The Treaty of Guadalupe Hidalgo was formally proclaimed on July 4, 1848, although the American army already had evacuated Mexico City.

CHAPTER THREE: CALIFORNIA GOLD

1. Porter, "Stone," 4; Passport Applications, 1795–1905, RG 59, General Records of the Department of State, National Archives.

2. Ibid.; Brian Holden Reid, *The Civil War and the Wars of the Nineteenth Century* (New York: Harper Collins, 1999), 30–31.

3. Ibid., 29–30.

4. Cullum, *Biographical Register*, 117.

5. Brevet Brigadier General Bennet Riley to Lieutenant Colonel W. G. Freeman, Monterey, California, February 28, 1850, in *Message from the President of the United*

States, in Answer to a Resolution of the Senate calling for further information in relation to the formation of a State Government in California; and also in relation to the condition of Civil Affairs in Oregon, 31st Congress, 1st Session, Executive Document no. 52, 69; Charles P. Stone to Lt. Col. J. W. Ripley, San Francisco, September 8, 1855, in *Correspondence Between Late Secretary of War and General Wool,* House Exec. Document, 35th Cong., 1st Sess., Ser. Vol. 956, Report 88, 1857–1858, 139–141; Josephine W. Cowell, *History of the Benicia Arsenal, Benicia, California, January 1851–December 1962* (Berkeley: Howell-North Books, 1963), 9.

6. *Regulations for the Government of the Ordnance Department* (Washington: Printed by Gideon, 1852), 3.

7. *Report of the Secretary of War, with Statements showing the contracts made under authority of that Department during the year 1851,* Senate Executive Document, 32nd Congress, 1st Session, Ex. Doc. no. 12, 15; Carl C. Cutler, *Queens of the Western Ocean: The Story of America's Mail and Passenger Sailing Lines* (Annapolis: United States Naval Institute, 1961), 385–386, 404, 406; January 13, February 2, 1851, Journal of Orders, Benicia Arsenal Log, Benicia Historical Museum Collection, Benicia, California; Cullum, *Biographical Register 2,* 220. An "artificer" was an enlisted man, skilled in mechanics and weaponry, who would repair and maintain small arms.

8. February 28, 1851, Journal of Orders, Benicia Arsenal Log; William T. Sherman, *Memoirs of General William T. Sherman,* 2 vols., 2nd Ed. (New York: D. Appleton and Company, 1904), vol. 1, 41.

9. T. Robinson Warren, *Dust and Foam; or, Three Oceans and Two Continents* (New York: Charles Scribner, 1849), 21–22.

10. Sherman, *Memoirs,* vol. 1.

11. April 25, May 19, 1851, Journal of Orders, Benicia Arsenal Log.

12. July 3, 1851, Journal of Orders, Benicia Arsenal Log; Cowell, *Benicia Arsenal,* 11; (San Francisco) *Daily Alta California,* August 16, 1851.

13. Stone to Ripley, September 8, 1855; Cowell, *Benicia Arsenal,* 11–12; Returns from United States Military Posts, 1800–1916, Ordnance Depot Near Benicia, August 1851, RG 94, Records of the Adjutant General's Office, National Archives.

14. Eugene Bandel, *Frontier Life in the Army, 1854–1861* (Glendale, CA: Arthur H. Clark Company, 1932), 301.

15. Stone to Ripley, September 8, 1855; Cowell, *Benicia Arsenal,* 12.

16. Robert Bruegmann, *Benicia: Portrait of an Early California Town; An Architectural History* (Fairfield, CA: James Stevenson, Publisher, 1997), 67–68; Cowell, *Benicia Arsenal,* 13; Returns from United States Military Posts, Ordnance Depot Near Benicia, September, 1851.

17. John Murray and James V. Swantek, *1813–1997, The Watervliet Arsenal: A Chronicle of the Nation's Oldest Arsenal* (Watervliet, NY: Watervliet Arsenal, 1998), 67; December 22, 1851, June 13, 1852, Journal of Orders, Benicia Arsenal Log; Richard Dillon, *Great Expectations: The Story of Benicia, California* (Benicia: Benicia Heritage Books Inc., 1980), 68; Cowell, *Benicia Arsenal,* 13–14.

18. Bruegmann, *Benicia Portrait,* 68; Dillon, *Great Expectations,* 68–69.

19. Report of the Colonel of Ordnance, in *Message from the President of the United States to the two Houses of Congress at the commencement of the second session of the Thirty-second Congress, December 6, 1852,* House Executive Documents, 32nd Cong., 2nd

Sess., Ser. Vol. 674, House Exec. Doc. 1, Pt. 2, 265–266; Hubert Howe Bancroft, *History of California, Vol. 1, 1542–1800* (San Francisco: History Company, Publishers, 1886), 698–700; Joseph J. Hawgood Jr., *Engineers at the Golden Gate* (San Francisco: US Army Corps of Engineers, San Francisco District, 1980), 23–24; Cowell, *Benicia Arsenal*, 15; Porter, "Stone," 5. In 1853, the army began construction of Fort Point on the site of the Castillo de San Joaquin, demolishing what was left of the original structure.

20. Stephen W. Sears, *George B. McClellan: The Young Napoleon* (New York: Ticknor & Fields, 1988), 37–38; Grant, *Personal Memoirs*, vol. 1, 161; Returns from United States Military Posts, Benicia, October, 1852, February, November, 1853, January, 1854; Cullum, *Biographical Register 2*, 281.

21. "The Arsenal at Benicia," *California Farmer and Journal of Useful Sciences* 3 (May 10, 1855): 147.

22. Sherman, *Memoirs*, vol. 1, 106.

23. September 8, 1853, January 27, 29, October 4, 1854, May 21, 1855, Journal of Orders, Benicia Arsenal Log.

24. Charles Wells Chapin, *Sketches of the Old Inhabitants of Old Springfield, and Other Citizens of the Present Century, and Its Historic Mansions of "Ye Olden Tyme"* (Springfield, MA: Springfield Printing and Binding Company, 1893), 146–147.

25. Robert W. Frazer, ed., *Mansfield on the Condition of the Western Forts, 1853–1854* (Norman: University of Oklahoma Press, 1963), 227; Cowell, *Benicia Arsenal*, 18; *Annual Report of the Quartermaster and Adjutant-General*, California State Assembly, Document no. 6, Session of 1856.

26. Special Orders no. 21, Headquarters Department of the Pacific, San Francisco, March 10, 1854, in *Instructions and Correspondence Between the Government and Major General Wool, in Regard to His Operations on the Coast of the Pacific, January 3, 1855*, Senate Ex. Doc. 16, Ser. Vol. 751, 33rd Cong., 2nd Sess., 13.

27. Porter, "Stone," 5; Frazer, *Condition of the Western Forts*, 156–157.

28. Ibid., 155–156; "The Powder Magazine at Benicia," *California Farmer and Journal of Useful Sciences* 3 (April 26, 1855): 131.

29. Grant, *Personal Memoirs*, 121.

30. Sherman, *Memoirs*, vol. 1, 101–102; Brooks D. Simpson, *Ulysses S. Grant: Triumph Over Adversity, 1822–1865* (Boston: Houghton Mifflin Company, 2000), 57; *California Pioneer Register and Index, 1542–1848: Including Inhabitants of California, 1769–1800 and List of Pioneers Extracted from the History of California by Hubert Howe Bancroft* (Baltimore: Regional Publishing Company, 1964), 145; John F. Marzalek, *Commander of All Lincoln's Armies: A Life of General Henry W. Halleck* (Cambridge, MA: Belknap Press of Harvard University Press, 2004), 55, 67–74, 76–77; Edward M. Coffman, *The Old Army: A Portrait of the American Army in Peacetime, 1784–1898* (New York: Oxford University Press, 1986), 59, 86.

31. Samuel Colville, comp., *Colville's San Francisco Directory, 1856, 1857*, vol. 1 (San Francisco: Commercial Steam Presses: Monson, Valentine & Co., 1856), 211; Certificate of Incorporation, San Francisco and Sacramento RailRoad Company, January 21, 1856, Incorporation Papers, California State Archives, Sacramento; Documents for the History of California: Papers of Thomas O. Larkin, vol. 9, pt. 3, no. 379, no. 427, Hubert Howe Bancroft Library, University of California, Berkeley; Harlan Hague and David J. Langum, *Thomas O. Larkin: A Life of*

Patriotism and Profit in Old California (Norman: University of Oklahoma Press, 1990), 220.

32. Dwight L. Clarke, *William Tecumseh Sherman: Gold Rush Banker* (San Francisco: California Historical Society, 1969), 93.

33. Ibid., 170.

34. Ira Cross, *Financing an Empire: History of Banking in California*, 4 vols. (Chicago: S. J. Clarke Publishing Company, 1927), vol. 1, 213–214.

35. Clarke, *Gold Rush Banker*, 231–233.

36. Ibid., 252, 254–255, 259–260, 264; Cross, *Financing an Empire*, vol. 1, 214.

37. Clarke, *Gold Rush Banker*, 255, 264–265.

38. Coffman, *Old Army*, 49; Cullum, *Biographical Register*, 117; Clarke, *Gold Rush Banker*, 262, 264; John S. Hittell, *The Resources of California: Comprising Agriculture, Mining, Geography, Commerce, Etc., and the Past and Future Development of the State* (San Francisco: A. Roman and Company, 1863), 334; Vital Records from *The Daily Evening Bulletin*, San Francisco, California, 1856, 1857, Volume I, 153, Compiled by Puerta de Oro Chapter, Genealogical Records Committee, California State Society, Daughters of the American Revolution, 1943, typescript in the California Room, California State Library, Sacramento.

39. Not long after his arrival in New York City, Sherman himself became caught up in the Panic of 1857, which resulted in the failure of Lucas, Turner, and his own unemployment. Stephen E. Woodworth, *Sherman* (New York: Palgrave Macmillan, 2009), 24–25.

40. Clarke, *Gold Rush Banker*, 265.

CHAPTER FOUR: MEXICAN MISADVENTURE

1. *Diccionario Porrua de Historia, Biografia y Geografia de Mexico*, Segunda Edicion (Mexico, DF: Editorial Porrua, SA, 1965), 836; *Before the Joint Commission of the United States and Mexico in the Matter of the Claim of the Lower California Company, Memorial and Exhibits, Sonora Survey* (New York: Evening Post Steam Presses, 1870), 1–3.

2. Sylvester Mowry, *Arizona and Sonora: The Geography, History, and Resources of the Silver Region of North America*, 3rd Ed. (New York: Harper & Brothers, 1864), 99–100.

3. (San Francisco) *Daily Alta California*, November 27, 1852; Laureano Calvo Berber, *Nociones de Historia de Sonora* (Mexico, DF: Libreria de Manuel Porrua, SA, 1958), 163–164; John Mayo, *Commerce and Contraband on Mexico's West Coast in the Era of Barron, Forbes & Co., 1821–1859* (New York: Peter Lang, 2006), 382–384; Stuart F. Voss, *On the Periphery of Nineteenth Century Mexico: Sonora and Sinaloa, 1810–1877* (Tucson: University of Arizona Press, 1982), 117–119; Rufus K. Wyllys, *The French in Sonora, 1850–1854: The Story of French Adventurers from California into Mexico* (Berkeley: University of California Press, 1932), 68–73; Joseph A. Stout Jr., *Schemers and Dreamers: Filibustering in Mexico, 1848–1921* (Fort Worth: Texas Christian University Press, 2002), 26–27.

4. *Claim of the Lower California Company (Sonora Survey)*, 24–30.

5. Ibid., 6, 32–37, 39–42; *Title Papers of the Lower California Company, to Lands, Etc., in the Territory of Lower California and in the States of Sonora and Sinaloa of the Republic of Mexico* (New York: Evening Post Steam Presses, 1870), 5.

6. Ibid., 7.

7. Charles D. Poston, "Building a State in Apache-Land," *Overland Monthly* 24 (July 1894): 88; (San Francisco) *Daily Alta California*, November 24, 1853; *Before the Joint Commission of the United States and Mexico in the Matter of the Claim of the Lower California Company, Memorial and Exhibits, Iturbide Grant* (New York: Evening Post Steam Presses, 1870), 1–2. At the time, a square league was considered to be the equivalent of about 4,428 acres. Stone may also have been involved in conducting some surveys for a railroad being promoted from Chihuahua to Guaymas. Alexander S. Taylor, *A Historical Summary of Baja California from Its Discovery to 1867* (Pasadena, CA: Socio-Technical Books, 1971), 120.

8. *Claim of the Lower California Company (Sonora Survey)*, 4; *Title Papers*, 69–70.

9. *Claim of the Lower California Company (Sonora Survey)*, 7.

10. Warren, *Dust and Foam*, 163–164; *Los Angeles Star*, June 5, 1858; Francisco R. Almada, *Diccionario de Historia, Geografia y Biografia Sonorenses* (Chihuahua, 1952), 572–583. For a detailed biography of Pesqueira, see Rodolfo F. Acuna, *Sonoran Strongman: Ignacio Pesqueira and His Times* (Tucson: University of Arizona Press, 1974).

11. *Title Papers*, 70; *Claim of the Lower California Company (Sonora Survey)*, 7–8; Acuna, *Sonoran Strongman*, 56–57.

12. Gaston Garcia Cantu, *Las Invasiones Norteamericanas en Mexico* (Mexico, DF: Fondo Cultural Economica, 1996), 178–179; *New York Evening Post*, May 29, 1858; *San Diego Herald*, May 2, 9, 1857; William O. Scroggs, *Filibusters and Financiers: The Story of William Walker and His Associates* (New York: Macmillan Company, 1916), 31–48.

13. Charles P. Stone, *Notes on the State of Sonora* (Washington, DC: Samuel Polkinghorn, Printer, 1860), 5.

14. Charles P. Stone to S. W. Inge, Guaymas, November 25, 1858, in Mowry, *Arizona and Sonora*, 102.

15. *Annual Report of the Board of Regents of the Smithsonian Institution, Showing Operations, Expenditures and Condition of the Institution for the Year 1860* (Washington, DC: George W. Bowman, Printer, 1861), 63.

16. *Claim of the Lower California Company (Sonora Survey)*, 55.

17. (San Francisco) *Daily Alta California*, October 13, 1858.

18. Bob [Robert] Whiting to Steve [?], Ft. Buchanan, Arizona, January 4, 1859, Whiting Family Letters, New York Historical Society Museum and Library, New York, NY; Mowry, *Arizona and Sonora*, 99.

19. David F. Long, *Gold Braid and Foreign Relations: Diplomatic Activities of U. S. Naval Officers, 1798–1898* (Annapolis: Naval Institute Press, 1988), 108; *Biographical Memoir of Charles Henry Davis, 1807–1877, Read Before the National Academy, April, 1896*, National Academy of Sciences Biographical Memoirs, http://www.nasonline.org/publications/biographicalmemoirs; Charles P. Stone to Secretary of State Lewis Cass, Guaymas, December 16, 1858, Dispatches from United States Consuls in Guaymas, Mexico, vol. 1, November 27, 1832–December 31, 1867, RG 59, Records of the Department of State, National Archives.

20. Charles P. Stone to Lewis Cass, Guaymas, December 23, 1858, Dispatches from United States Consuls in Guaymas.

21. Acuna, *Sonoran Strongman*, 58; Long, *Gold Braid*, 105; Stone provided an indication of what he envisioned for the future of Sonora in a November 1858 letter to an investor in the Jecker Contract when he wrote that the Commission's geologist, Thomas Antisell, would be particularly useful in identifying rich mines once the state had been annexed to the United States. *Claim of the Lower California Company (Sonora Survey)*, 67.
22. Charles P. Stone to Lewis Cass, Guaymas, January 17, 1859, Dispatches from United States Consuls in Guaymas.
23. Charles P. Stone to Lewis Cass, Guaymas, February 21, 1859, Robert Rose to Lewis Cass, February 4, 1859, Dispatches from United States Consuls in Guaymas.
24. *Claim of the Lower California Company (Sonora Survey)*, 66; Whiting to Steve, Ft. Buchanan, Arizona, January 4, 1859, Whiting Letters, New York Historical Society.
25. The Exequatur is a document issued by the head of state of a government to a foreign consul recognizing that person as an official representative and guaranteeing his or her certain rights and privileges.
26. Stone to Cass, Guaymas, January 17, February 21, 1859, Charles P. Stone to Captain Kelly, Guaymas, February 21, 1859, Robert Rose to Lewis Cass, Guaymas, January 17, February 14, March 31, May 27, 1859, Dispatches from United States Consuls in Guaymas; *The Works of James Buchanan Comprising His Speeches, State Papers, and Private Correspondence*, Ed. John Bassett Moore, vol. 10 (New York: Antiquarian Press Ltd., 1960), 256–257; Acuna, *Sonoran Strongman*, 59.
27. Acuna, *Sonoran Strongman*, 59.
28. *Claim of the Lower California Company (Sonora Survey)*, 76; Charles P. Stone to J. B. Jecker & Co., May 27, 1859, Dispatches from United States Ministers to Mexico, vol. 1, 1823–1906, RG 59, Records of the Department of State, National Archives; Michael James Box, *Capt. James Box's Adventures and Explorations in New and Old Mexico* (New York: Derby & Jackson, Publishers, 1861), 329.
29. Charles P. Stone to the Secretary of State, Washington, DC, August 25, 1859, in *Messages from the President of the United States to the Two Houses of Congress at the Commencement of the First Session of the Thirty-Sixth Congress*, Senate Exec. Doc. no. 2, 36th Cong., 1st Sess., vol. 1 (Washington, DC: George W. Bowman, Printer, 1860), 38–39; Tubac, *Weekly Arizonian*, June 16, 1859.
30. Ures, Sonora, *La Estrella de Occidente*, November 18, 1859; W. D. Porter to Capt. R. S. Ewell, Guaymas, November 11, 1859, RG 94, Letters Received by the Adjutant General (Main Series), 1822–1860, National Archives; Acuna, *Sonoran Strongman*, 61–62.
31. Ignacio Pesqueira to Farrelly Alden, Guaymas, November 1, November 2, November 3, 1859, Farrelly Alden to Capt. W. D. Porter, Guaymas, November 18, November 21, 1859, Farrelly Alden to Prefect Thomas Robinson, Guaymas, November 19, 1859, Proclamation of Thomas Robinson, Guaymas, November 21, 1859, Dispatches from United States Consuls in Guaymas; Acuna, *Sonoran Strongman*, 63.
32. Charles P. Stone to Secretary of War John B. Floyd, Washington, DC, September 28, 1859, Col. S. Cooper to Major E. D. Townsend, Ft. Monroe, Virginia, October 7, 1859, Letters Received by the Adjutant General (Main Series).

33. W. D. Porter to Capt. R. S. Ewell, Guaymas, November 11, 1859, R. S. Ewell to Col. S. Cooper, Ft. Buchanan, Arizona, November 30, 1859, Letters Received by the Adjutant General (Main Series); Farrelly Alden to Lewis Cass, Guaymas, November 26, 1859, Dispatches from United States Consuls in Guaymas; Pfanz, *Ewell*, 114–115; Acuna, *Sonoran Strongman*, 63.

34. *Claim of the Lower California Company (Iturbide Grant)*, 5; *Claim of the Lower California Company (Sonora Survey)*, 10; *New York Times*, January 13, 1860; William Turrentine Jackson, *Wagon Roads West: A Study of Federal Road Surveys and Construction in the Trans-Mississippi West, 1846–1869* (New Haven: Yale University Press, 1965), 232; Farrelly Alden to Lewis Cass, Guaymas, March 18, 1860, Dispatches from United States Consuls in Guaymas.

35. Caleb Cushing, *Contract of the Mexican Government for the Survey of the Public Lands of the State of Sonora* (Washington, DC, 1860), 39.

36. *New York Times*, December 15, 1859, January 13, 1860.

37. (San Francisco) *Daily Evening Bulletin*, November 19, 1859.

38. *Claim of the Lower California Company (Sonora Survey)*, 11–13; Caesar Count Corti, *Maximilian and Charlotte of Mexico*, vol. 1 (New York: Alfred A. Knopf, 1928), 131–132; Pablo L. Martinez, *A History of Lower California*, trans. Ethel Duffy Turner (Mexico, DF: Editorial Baja California, 1960), 389–394; Angela Moyano Pahissa, *California y sus Relaciones con Baja California* (Mexico, DF: Fondo del Cultural Economica, 1983), 64–65; Adrian Valades, *Historia dela Baja California, 1850–1880* (Mexico, DF: Universidad Nacional Autonoma de Mexico, 1974), 194–196.

39. Almada, *Diccionario*, 580–583; Pesqueira, *Sonoran Strongman*, 134–135.

CHAPTER FIVE: DEFENDING THE CAPITAL

1. Margaret Leech, *Reveille in Washington, 1860–1865* (New York: Harper & Brothers, 1941), 1–4, 13; John S. D. Eisenhower, *Agent of Destiny: The Life and Times of General Winfield Scott* (New York: Free Press, 1997), 348.

2. Russell H. Beatie, *Army of the Potomac: Birth of Command, November 1860–September 1861* (Boston: Da Capo Press, 2002), 8, 19; Eisenhower, *Agent*, 344, 381–382; Leech, *Reveille*, 2, 28; Charles P. Stone, "Washington on the Eve of the War," *Battles and Leaders of the Civil War*, vol. 1 (New York: Century Company, 1887–1888), 7; Edwin C. Fishel, *The Secret War for the Union: The Untold Story of Military Intelligence in the Civil War* (Boston: Houghton Mifflin Company, 1996), 12.

3. Stone, "Washington," 9–11.

4. Allan Peskin, *Winfield Scott and the Profession of Arms* (Kent, OH: Kent State University Press, 2003), 244-245.

5. Ibid., 14–15; Beatie, *Potomac*, 21.

6. Stone, "Washington," 11–12; Leech, *Reveille*, 28.

7. Stone, "Washington," 15–17.

8. Ibid., 17.

9. Ibid.; Beatie, *Potomac*, 23.

10. Stone, "Washington," 13; David Detzer, *Dissonance: The Turbulent Days Between Fort Sumter and Bull Run* (Orlando: Harcourt Inc., 2006), 12–13; Leech, *Reveille*, 29.

11. Stone, "Washington," 18–20; John G. Nicolay and John Hay, "The President Elect in Springfield," *Century* 35 (November 1887): 71.

12. Stone, "Washington," 22–23; Eisenhower, *Agent*, 351–352.

13. Ernest B. Furgurson, *Freedom Rising: Washington in the Civil War* (New York: Alfred A. Knopf, 2004), 39; Leech, *Reveille*, 31–32.

14. Leech, *Reveille*, 33–34.

15. Winfield Scott to William Seward, Washington, February 21, 1861; Memorandum of Charles P. Stone, Washington, February 21, 1861, Abraham Lincoln Papers at the Library of Congress; Michael J. Kline, *The Baltimore Plot: The First Conspiracy to Assassinate Abraham Lincoln* (Yardley, PA: Westholme Publishing, 2008), 194–196.

16. Stone, "Washington," 23–24; Kline, *Baltimore Plot*, 197.

17. Furgurson, *Freedom*, 59.

18. Stone, "Washington," 21, 24–25; Leech, *Reveille*, 41; Kline, *Baltimore Plot*, 333; Fishel, *Secret War*, 15.

19. Stone, "Washington," 25; Leech, *Reveille*, 43–45; Kline, *Baltimore Plot*, 28–29; Ben Perley Poore, *Pereley's Reminiscence of Sixty Years in the National Metropolis*, vol. 2 (Philadelphia: Hubbard Brothers, Publishers, 1888), 58; Benson J. Lossing, *Pictorial History of the Civil War of the United States of America*, vol. 1 (Philadelphia: George W. Childs, Publisher, 1866), 288.

20. Charles P. Stone, "Washington in March and April, 1861," *Magazine of American History* 14 (July 1885): 1, 3–6; Winfield Scott to Abraham Lincoln, Washington, April 5, 9, 10, 11, 1861, Abraham Lincoln Papers at the Library of Congress; Leech, *Reveille*, 54; Kenneth J. Winkle, *Lincoln's Citadel: The Civil War in Washington, DC* (New York: W. W. Norton & Company, 2013), 144–145.

21. Stone, "March and April," 8.

22. Ibid., 11; Cullum, *Biographical Register*, 117; John Lockwood and Charles Lockwood, *The Siege of Washington: The Untold Story of the Twelve Days that Shook the Union* (New York: Oxford University Press, 2011), 64.

23. Stone, "March and April," 8–11.

24. Ibid., 12–14.

25. Ibid., 14–15; Richard B. Kielbowicz, "The Telegraph, Censorship, and Politics at the Outset of the Civil War," *Civil War History* 40 (June 1994): 97–98; Detzer, *Dissonance*, 208–209.

26. Stone, "March and April," 12.

27. Charles P. Stone, "A Dinner with General Scott," *Magazine of American History* 16 (January–June 1884): 528–529.

28. Ibid., 530–531.

29. Ibid.

30. Ibid., 532.

31. Ibid.; Lockwood and Lockwood, *Siegel*, 170–171.

32. *The War of the Rebellion: A Compilation of the Official Records of the Union and Confederate Armies*, *OR*, series 1, vol. 2 (Washington: Government Printing Office, 1880), 602.

33. Scott and Stone did not know it at the time, but Virginia lacked the warships to impose an effective blockade on the Potomac River.

34. Stone, "March and April," 18–19.

35. Ibid., 20.

36. Ibid.; Winfield Scott to Abraham Lincoln, Washington, April 24, 1861, Abraham Lincoln Papers at the Library of Congress; William Marvel, *Mr. Lincoln Goes to War* (Boston: Houghton Mifflin, 2006), 41.

37. Stone, "March and April," 24; Lockwood and Lockwood, *Siege*, 197; *OR*, series 1, vol. 2, 600.

38. Leech, *Reveille*, 65; Lockwood and Lockwood, *Siege*, 229–230; *OR*, series 1, vol. 51, 344; Stone, "March and April," 24.

CHAPTER SIX: GENERAL STONE

1. Cullum, *Biographical Register*, 118.

2. William Marvel, *Mr. Lincoln*, 73–76; Frank J. Welcher, *The Union Army, 1861–1865: Organization and Operations, Volume I, The Eastern Theater* (Bloomington: Indiana University Press, 1989), 71; Charles P. Stone to Benson Lossing, Goochland County, Virginia, November 5, 1866, William L. Clements Library, University of Michigan, Ann Arbor.

3. General Orders no. 1, Department of Northeastern Virginia, May 28, 1861, *OR*, series 1, vol. 51, 389–390; E. D. Townsend to C. P. Stone, Washington, DC, June 8, 1861, *OR*, series 1, vol. 2, 104; Stone to Lossing, November 5, 1866, Clements Library.

4. George A. Hussey, *History of the Ninth Regiment, NYSM–NGSNY (Eighty-Third NY Volunteers), 1845–1888* (New York: Veterans of the Regiment, 1888), 47; John Wesley Jaques, *Three Years' Campaign of the Ninth NYSM, During the Southern Rebellion* (New York: Hilton & Co., Publishers, 1865), 16; Timothy R. Snyder, "Securing the Potomac: Colonel Charles P. Stone and the Rockville Expedition, June–July 1861," *Catoctin History*, no. 11 (2009): 9–10.

5. Marvel, *Mr. Lincoln*, 94; Jaques, *Three Years' Campaign*, 17–20; Snyder, "Securing the Potomac," 11.

6. Charles P. Stone to E. D. Townsend, Poolesville, June 17, 1861, *OR*, series 1, vol. 2, 109.

7. Charles P. Stone to E. D. Townsend, Poolesville, June 17, 1861, *OR*, series 1, vol. 2, 95.

8. Charles P. Stone to E. D. Townsend, Poolesville, June 18, 1861, *OR*, series 1, vol. 2, 111–112.

9. Eppa Hunton, *Autobiography of Eppa Hunton* (Richmond, VA: William Byrd Press, 1933), 28; Snyder, "Securing the Potomac," 15.

10. E. D. Townsend to Charles P. Stone, Washington, DC, June 8, 1861, Charles P. Stone to E. D. Townsend, Poolesville, June 15, June 16, June 18, 1861, *OR*, series 1, vol. 2, 104–105, 108, 109, 111–112.

11. Marvel, *Mr. Lincoln*, 97–98; Charles P. Stone to E. D. Townsend, Poolesville, June 22, 23, 1861, *OR*, series 1, vol. 2, 114–115.

12. Charles P. Stone to E. D. Townsend, Poolesville, June 20, 24, 1861, *OR*, series 1, vol. 2, 113, 115–116; *Washington Evening Star*, June 25, 1861.

13. Special Orders No. 109, Washington, DC, June 30, 1861, *OR*, series 1, vol. 51, 407.

14. Hussey, *Ninth Regiment*, 53–54; Jaques, *Three Years' Campaign*, 24–25; *Washington Evening Star*, July 8, 1861.

15. *Report of the Joint Committee on the Conduct of the War*, 37th Congress, 3rd Session (Washington: Government Printing Office, 1863), 74; Charles P. Stone to E. D. Townsend, Point of Rocks, July 4, 1861; F. J. Porter to Charles P. Stone, Martinsburg, Virginia, July 4, 1861, *OR*, series 1, vol. 2, 120–122.

16. Charles P. Stone to Fitz John Porter, Opposite Harper's Ferry, July 5, 1861, F. J. Porter to Charles P. Stone, Martinsburg, July 5, 1861, E. D. Townsend to Charles P. Stone, Washington, DC, July 6, 1861, *OR*, series 1, vol. 51, 410–412.

17. Welcher, *Union Army*, 76, 77; *Joint Committee*, 84–86.

18. *Joint Committee*, 131; Journal of Major Abner Doubleday from May–July 1861, Janet B. Hewett, *Supplement to the Official Records of the Union and Confederate Armies* (Wilmington, NC: Broadfoot Publishing Company, 1994), Part I, Reports, Volume 1, Serial no. 1, 199; Welcher, *Union Army*, 77.

19. Fred Harvey Harrington, *Fighting Politician: Major General N. P. Banks* (Philadelphia: University of Pennsylvania Press, 1948), 62.

20. Hewett, *Supplement*, 199; E. D. Townsend to C. P. Stone, Washington, August 3, 1861, Generals' Papers, Entry 159, Stone, Charles P., RG 94, National Archives; General Orders no. 36, Department of the Shenandoah, Sandy Hook, July 29, 1861, Special Orders no. 10, Division of the Potomac, Washington, DC, August 4, 1861, *OR*, series 1, vol. 51, 428–429, 434–435; Welcher, *Union Army*, 77, 86–87.

21. Cullum, *Biographical Register*, 118; Charles W. Sanford to Winfield Scott, Washington, August 6, 1861, Stone Entry, Generals' Papers.

22. G. B. McClellan to C. P. Stone, Washington, August 18, 1861, Stephen W. Sears, ed., *The Civil War Papers of George B. McClellan: Selected Correspondence, 1860–1865* (New York: Ticknor & Fields, 1989), 86–87; Beatie, *Potomac*, 466–467.

23. Charles P. Stone to Major S. Williams, Seneca, MD, August 22, 1861, *OR*, series 1, vol. 5, 578–579; Marvel, *Mr. Lincoln*, 200–201.

24. William F. Howard, *The Battle of Ball's Bluff: The Leesburg Affair, October 21, 1861* (Lynchburg, VA: H. E. Howard, 1994), 4; Charles P. Stone to Captain Robert Williams, Poolesville, September 2, 1861, *OR*, series 1, vol. 51, 466–467; Andrew E. Ford, *History of the Fifteenth Massachusetts Volunteer Infantry in the Civil War, 1861–1864* (Clinton: Press of W. J. Coulter, 1898), 54.

25. Huton, *Autobiography*, 26.

26. Byron Farwell, *Ball's Bluff: A Small Battle and Its Long Shadow* (McLean, VA: EPM Publications Inc., 1990), 40–44.

27. Frank Moore, ed., *Rebellion Record: A Diary of American Events, With Documents, Narratives, Illustrative Incidents, Poetry, Etc.* (New York: G. Putnam, 1862), vol. 4, Documents, 11; Ford, *Fifteenth Massachusetts*, 58; Farwell, *Ball's Bluff*, 60; Marvel, *Mr. Lincoln*, 207.

28. George B. McClellan, *McClellan's Own Story: The War for the Union, the Soldiers Who Fought It, the Civilians Who Directed It, and His Relations to It and Them* (New York: Charles L. Webster & Company, 1887), 181–182; Farwell, *Ball's Bluff*, 64–65; Kim Bernard Holien, *The Battle at Ball's Bluff* (Orange, VA: Publisher's Press, 1995), 20–22; Marvel, *Mr. Lincoln*, 218, 220; Ethan S. Rafuse, *McClellan's War: The Failure of Moderation in the Struggle for the Union* (Bloomington: Indiana University Press, 2005), 136–137, 139.

29. *OR*, series 1, vol. 5, 293.

30. Farwell, *Ball's Bluff*, 65–66; Holien, *Battle*, 24–26; James A. Morgan III, *A Little Short of Boats: The Fights at Ball's Bluff and Edward's Ferry, October 21–22, 1861* (Ft. Mitchell, KY: Ironclad Publishing Inc., 2004), 32.
31. Howard, *Leesburg Affair*, 12; Morgan, *Boats*, 33–34; *Joint Committee*, 277; Moore, *Rebellion Record*, vol. 3, Documents, 98.

CHAPTER SEVEN: TRIPPED UP BY CIRCUMSTANCES

1. *Joint Committee*, 10, 267, 277; *OR*, series 1, vol. 5, 294; Holein, *The Battle*, 27; Farwell, *Ball's Bluff*, 67–68.
2. *Joint* Committee, 268; *OR*, series 1, vol. 5, 294; Farwell, *Ball's Bluff*, 69–70; Howard, *Leesburg Affair*, 13; Morgan, *Boats*, 59–60.
3. Morgan, *Boats*, 36, 38–39.
4. Ibid., 40.
5. Farwell, *Ball's Bluff*, 77.
6. *Joint Committee*, 252; Holein, *The Battle*, 28–30; Donald R. Jermann, *Civil War Battlefield Orders Gone Awry: The Written Word and Its Consequences in Thirteen Engagements* (Jefferson, NC: McFarland & Company, Publishers, 2012), 17.
7. Morgan, *Boats*, 68–69.
8. *Joint Committee*, 268–269; Morgan, *Boats*, 69.
9. *Joint Committee*, 275; Morgan, *Boats*, 69–70; *OR*, series1, vol. 5, 303.
10. Telegrams, Stone to McClellan, Edward's Ferry, October 21, 1861, McClellan to Stone, Headquarters, October 21, 1861, *OR*, series 1, vol. 51, 498.
11. *Joint Committee*, 376.
12. Ibid., 269, 377; Morgan, *Boats*, 94.
13. Telegram, Stone to McClellan, Edward's Ferry, October 21, 1861, *OR*, series 1, vol. 5, 33.
14. McClellan, *Story*, 183; Telegram, McClellan to Stone, Headquarters, October 21, 1861, *OR*, series 1, vol. 51, 499.
15. Charles P. Stone to E. D. Baker, Edward's Ferry, October 21, *OR*, series 1, vol. 5, 303; Telegram, Stone to McClellan, Edward's Ferry, October 21, 1861, *OR*, series 1, vol. 51, 499.
16. E. D. Baker to Stone, Conrad's Ferry, October 21, 1861, *OR*. series 1, vol. 51, 502.
17. *Joint Committee*, 488; Albert James Myer, "Operations and Duties of the Signal Department of the Army, 1860–1865," Hewett, *Supplement to the OR*, series 10, vol. 10, 315–316; Morgan, *Boats*, 127.
18. *Joint Committee*, 488; Telegram, Stone to McClellan, Edward's Ferry, October 21, 1861; *OR*, series 1, vol. 5, 33–34.
19. Telegram, Stone to McClellan, Edward's Ferry, October 21, 1861, *OR*, series 1, vol. 5, 34; Telegram, Stone to Banks, Edward's Ferry, October 21, 1861, *OR*, series 1, vol. 51, 501; *OR*, series 1, vol. 5, 298; Morgan, *Boats*, 128.
20. *Joint Committee*, 308; Farwell, *Ball's Bluff*, 92.
21. *OR*, series 1, vol. 5, 321; Farwell, *Ball's Bluff*, 93.
22. Morgan, *Boats*, 111, 144–145; *OR*, series 1, vol. 5, 328.
23. *OR*, series 1, vol. 5, 322; Farwell, *Ball's Bluff*, 105–106; Morgan, *Boats*, 149, 163–164; Holein, *The Battle*, 66; *Joint Committee*, 410.
24. *OR*, series 1, vol. 5, 311; Morgan, *Boats*, 170–173.

25. Ibid., 130–131; *OR*, series 1, vol. 5, 298.

26. Telegram, Stone to McClellan, Edward's Ferry, October 21, 1861, McClellan, *Story*, 185.

27. Geoffrey Perret, *Lincoln's War: The Untold Story of America's Greatest President as Commander in Chief* (New York: Random House, 2004), 95–96; Doris Kearns Goodwin, *Team of Rivals: The Political Genius of Abraham Lincoln* (New York: Simon & Schuster, 2005), 381; Blair and Tarshis, *Baker*, 156–157.

28. Telegram, Stone to Banks, Edward's Ferry, October 21, 1861, *OR*, series 1, vol. 51, 502.

29. *Joint Committee*, 487; *OR*, series 1, vol. 5, 298.

30. *Joint Committee*, 487.

31. Telegram, Stone to McClellan, Edward's Ferry, October 21, 1861, McClellan, *Story*, 185.

32. Telegrams, McClellan to Stone, Headquarters, October 21, 1861, Stone to McClellan, Edward's Ferry, October 21, 1861, *OR*, series 1, vol. 51, 500.

33. Telegrams, A. Lincoln to Officer in Command at Poolesville, Washington, October 21, 1861, Stone to Lincoln, Edward's Ferry, October 21, 1861, *OR*, series 1, vol. 51, 498.

34. Telegrams, A. Lincoln to Officer in Command at Poolesville, Washington, October 21, 1861, Stone to Lincoln, Edward's Ferry, October 21, 1861, *OR*, series 1, vol. 51, 498.

35. *OR*, series 1, vol. 5, 333–334; Morgan, *Boats*, 193.

36. *OR*, series 1, vol. 5, 313.

37. *OR*, series 1, vol. 5, 35, 330–332.

38. Stone to Lossing, November 5, 1866, Clements Library.

39. *Joint Committee*, 268.

40. Ibid., 256.

Chapter Eight: The Committee

1. F. Stansbury Haydon, *Military Ballooning During the Early Civil War* (Baltimore: Johns Hopkins University Press, [1941] 2000), 360–364; Charles P. Stone to Professor Lowe, Poolesville, January 20, January 22, 1862, *OR*, series 3, vol. 3, 269.

2. Russell H. Beatie, *The Army of the Potomac, Volume II: McClellan Takes Command, September 1861–February 1862* (Boston: Da Capo Press, 2004), 419; Marvel, *Mr. Lincoln*, 268.

3. Charles P. Stone to Brigadier General S. Williams, Poolesville, October 29, 1861, *OR*, series 1, vol. 5, 293–299.

4. Telegram, McClellan to Division Commanders, Army of the Potomac, Poolesville, October 24, 1861, Sears, ed., *Civil War Papers*, 111.

5. George B. McClellan to Mary Ellen McClellan, Washington, October 25, 1861, Sears, ed., *Civil War Papers*, 111.

6. Ibid.

7. Blair and Tarshis, *Baker*, 157–159; Perret, *Lincoln's War*, 96–97.

8. *New York Times*, October 31, 1861.

9. Charles P. Stone to Brigadier General S. Williams, Poolesville, November 2, 1861, *OR*, series 1, vol. 5, 300–301.

10. Charles P. Stone to Lt. Colonel Hardie, Poolesville, December 2, 1861, *OR*, series 1, vol. 5, 307.

11. Beatie, *McClellan*, 407.
12. Thomas Drew to Lieutenant Colonel Palfrey, Boston, December 9, 1861, *OR*, series 2, vol. 1, 787–788.
13. Charles P. Stone to Brigadier General S. Williams, Poolesville, December 15, 1861, *OR*, series 2, vol. 1, 786–787.
14. Bruce Tap, *Over Lincoln's Shoulder: The Committee on the Conduct of the War* (Lawrence: University Press of Kansas, 1998), 58; T. Harry Williams, *Lincoln and the Radicals* (Madison: University of Wisconsin Press, 1965), 95.
15. Alfred R. Conkling, *The Life and Letters of Roscoe Conkling, Orator, Statesman, Advocate* (New York: Charles L. Webster & Company, 1889), 140–141; Matthew Phillips, "Bungled River Crossing," *Military History Magazine's Great Battles* (July 1991): 40.
16. Tap, *Shoulder*, 21–24; Hans L. Trefousse, "The Joint Committee on the Conduct of the War: A Reassessment," *Civil War History* 10 (March 1964), 5–19.
17. Conkling, *Life and Letters*, 147.
18. Marvel, *Mr. Lincoln*, 268; Stone Testimony, *Joint Committee*, 73–75, 265–278.
19. Stone Testimony, *Joint Committee*, 279.
20. Ibid., 280.
21. Ibid., 279–280.
22. James G. Blaine, *Twenty Years of Congress: From Lincoln to Garfield, with a Review of the Events Which Led to the Political Revolution of 1860* (Norwich, CT: Henry Bill Publishing Company, 1884), 383; Williams, *Radicals*, 96; *Joint Committee*, 295.
23. *Joint Committee*, 362.
24. Tap, *Shoulder*, 67; Order No. ___, Washington, January 28, 1862, *Joint Committee*, 502.
25. Richard B. Irwin, "Ball's Bluff and the Arrest of General Stone," *Battles and Leaders of the Civil War*, vol. 2 (New York: Century Company, 1884–1887), 133; Tap, *Shoulder*, 67; Williams, *Radicals*, 98.
26. Beatie, *McClellan*, 516; Stone to Lossing, Clements Library.
27. Stone Testimony, *Joint Committee*, 426–429; Blaine, *Twenty Years*, 384–385.
28. Stone Testimony, *Joint Committee*, 429.
29. Ibid., 430–433.

CHAPTER NINE: NADIR

1. Goodwin, *Rivals*, 415–417; Perley, *Reminiscences*, vol. 2, 115–120.
2. Stone to Lossing, November 5, 1866, Clements Library; Stephen W. Sears, *Controversies and Commanders: Dispatches from the Army of the Potomac* (Boston: Houghton Mifflin, 1999), 29–30.
3. Stone to Lossing, November 5, 1866, Clements Library; Sears, *Controversies*, 30.
4. Stone to Lossing, November 5, 1866, Clements Library.
5. Sears, *Controversies*, 42; George B. McClellan to Brigadier General Andrew Porter, Washington, DC, February 8, 1862, Sears, ed., *Civil War Papers*, 173.
6. Stone to Lossing, November 5, 1866, Clements Library.
7. Ibid.
8. Ibid.; George B. McClellan to Lieutenant Colonel M. Burke, Washington, DC, February 8, 1862, Generals' Papers, Entry 159, National Archives; *New York Times*, September 24, 1861.

9. Lonnie R. Speer, *Portals to Hell: Military Prisons of the Civil War* (Mechanicsburg, PA: Stackpole, 1997), 35.

10. First Lieutenant Charles O. Wood to Martin M. Burke, Fort Lafayette, April 9, 1862, *OR*, series 2, vol. 3, 440–441; Speer, *Portals*, 36.

11. Stone to Lossing, November 5, 1866, Clements Library; *New York Times*, September 24, 1861.

12. Charles P. Stone to Brigadier General L. Thomas, Washington, DC, September 22, 1862, *OR*, series 1, vol. 5, 329; Stone to Lossing, November 5, 1866, Clements Library; Beatie, *McClellan*, 527.

13. Charles O. Wood to M. Burke, Fort Lafayette, April 11, 1862; Joseph H. Bradley to Edwin M. Stanton, Washington, DC, February 16, 1862; P. H. Watson to Joseph H. Bradley, Washington, DC, February 20, 1862, *OR*, series 2, vol. 3, 265, 287, 442; George B. McClellan to Martin Burke, Washington, DC, February 16, 1862, A. G. Clary to General McClellan, Fort Hamilton, February 24, 1862, Generals' Papers, Entry 159, National Archives; Webb Garrison Jr., *Strange Battles of the Civil War* (New York: Bristol Park Books, 2001), 122.

14. *New York Times*, September 24, 1861; L. Thomas to Martin Burke, Washington, DC, February 20, 1862, Generals' Papers, Entry 159, National Archives; L. Thomas to Martin Burke, Washington, DC, March 12, 1862, D. Lynde to Martin Burke, Fort Hamilton, March 15, 1862, War Department Memorandum, Washington, DC, March 28, 1862, *OR*, series 2, vol. 3, 373, 380–381, 407.

15. Parole of Brigadier General Charles P. Stone, March 30, 1862, E. D. Townsend to Martin Burke, Washington, DC, April 7, 1862, Generals' Papers, Entry 159, National Archives.

16. Sears, *Controversies*, 44; Beatie, *McClellan*, 528.

17. Henry M. Parker to E. Stanton, Washington, DC, February 20, 1862, *OR*, series 2, vol. 3, 292–294; Beatie, *McClellan*, 528, T. Harry Williams, "Investigation, 1862," *American Heritage* 6 (December 1954): 4.

18. George B. McClellan to Martin Burke, Washington, DC, March 11, 1862, Generals' Papers, Entry 159, National Archives; *Revised Regulations for the Army of the United States, 1861* (Philadelphia: J. B. Lippincott & Co., Publishers, 1862), 512; H. Donald Winkler, *Civil War Goats and Scapegoats* (Nashville: Cumberland House, 2008), 54.

19. Henry M. Parker to L. Thomas, Washington, DC, April 20, 1862, *OR*, series 2, vol. 3, 466; Garrison, *Strange Battles*, 122.

20. *Speech of Hon. J. A. McDougall of California on the Arrest of Gen. Stone and the Rights of the Soldier and Citizen* (Washington, DC: L. Towers & Co., 1862), 1–2; Oscar T. Shuck, *Bench and Bar in California: History, Anecdotes, Reminiscences* (San Francisco: Occident Printing House, 1889), 357–360.

21. *Speech of Hon. J. A. McDougall*, 10–11; US Congress, Senate, Senator McDougall Speaking for the Resolution Concerning Brigadier General Charles P. Stone, 37th Congress, 2nd Session, April 15, 1862, *Congressional Globe* (1862), pt. 2: 1666.

22. US Congress, Senate, Senator Wade Speaking Against the Resolution Concerning Brigadier General Charles P. Stone, 37th Congress, 2nd Session, April 15, 1862, *Congressional Globe* (1862), pt. 2: 1667, 1668.

23. US Congress, Senate, Resolution Concerning Brigadier General Charles P. Stone, April 21, 1862, 37th Congress, 2nd Session, April 21, 1862, *Congressional Globe* (1862), pt. 2: 1732, 1742.

24. Abraham Lincoln to the Senate of the United States, Washington, DC, May 1, 1862, *The Collected Works of Abraham Lincoln*, vol. 5, ed. Roy P. Basler (New Brunswick: Rutgers University Press, 1953), 204.

25. Blaine, *Twenty Years*, 390–391; Stone to Lossing, November 5, 1866, Clements Library; E. D. Townsend to Martin Burke, Washington, DC, August 16, 1862, Generals' Papers, Entry 159, National Archives.

26. George B. McClellan to E. M. Stanton, Washington, DC, September 7, 1862, *OR*, series 1, vol. 5, 342; Stone to Lossing, November 5, 1866, Clements Library; Winkler, *Scapegoats*, 54–55; War Department, Adjutant General's Office, Washington, DC, January 26, 1863, Generals' Papers, Entry 159, National Archives.

27. H. W. Halleck to Charles P. Stone, Washington, DC, September 30, 1862, *OR*, series 1, vol. 5, 344; Stone to Lossing, November 5, 1866, Clements Library; Leech, *Reveille*, 184.

28. Blaine, *Twenty Years*, 393–394.

29. *Joint Committee*, 17–18.

30. Excerpts from the Journal of Major General Samuel Peter Heintzleman, US Army, January 26–June 3, 1863, Hewett, *Supplement*, vol. 4, Serial 4, 468; Curt Anders, *Disaster in Damp Sand: The Red River Expedition* (Carmel: Guild Press of Indiana, 1997), 91; Harrington, *Politician*, 118.

31. James Hollandsworth, *Pretense of Glory: The Life of Nathaniel P. Banks* (Baton Rouge: Louisiana State University Press, 1998), 83–84; Harrington, *Politician*, 92–93; John D. Winters, *The Civil War in Louisiana* (Baton Rouge: Louisiana State University Press, 1963), 147, 206.

32. A. J. H. Duganne, *Camps and Prisons: Twenty Months in the Department of the Gulf* (New York: J. P. Robens, Publishers, 1865), 19–20.

33. Charles P. Stone to N. P. Banks, Port Hudson, June 14, 27, 1863, N. P. Banks to Major General Frank Gardner, Port Hudson, July 8, 1863, Articles of Capitulation, Port Hudson, Louisiana, July 8, 1863, *OR*, series 1, vol. 26, pt. 1, 53–54, 102; Edward Cunningham, *The Port Hudson Campaign, 1862–1863* (Baton Rouge: Louisiana State University Press, 1963), 117–119.

34. [Charles P. Stone] to Thomas J. Buffington, Port Hudson, July 17, 1863, [Charles P. Stone] to N. P. Banks, Port Hudson, July 15, 1863, [Charles P. Stone] to Major General Ulysses S. Grant, Port Hudson, July 16, 1863, Richard B. Irwin to Brigadier General George L. Andrews, Port Hudson, July 14, 1863, *OR*, series 1, vol. 26, pt. 1, 641–643; Cunningham, *Port Hudson*, 120.

35. [Charles P. Stone] to N. P. Banks, Port Hudson, July 20, 1863, *OR*, series 1, vol. 26, pt. 1, 648; Winters, *Louisiana*, 279; Cunningham, *Port Hudson*, 212.

36. Charles P. Stone to N. P. Banks, Port Hudson, July 21, 1863, *OR*, series 1, vol. 26, pt. 1, 649–650; Harrington, *Politician*, 113.

37. C. P. S[tone] Memorandum for General [Banks], September 26, 1863, General Orders No. 70, Department of the Gulf, September 28, 1863, *OR*, series 1, vol. 26, pt. 1, 737, 741.

38. General Orders No. 54, Department of the Gulf, New Orleans, July 29, 1863, *OR*, series 1, vol. 26, pt. 1, 660.

39. Special Orders No. 58, New Orleans, July 28, 1863, *OR*, series 1, vol. 26, pt. 1, 657–658.

40. C. P. Stone to E. O. C. Ord, New Orleans, September 17, 1863, Edward Otho Cresap Ord Papers, Bancroft Library, University of California, Berkeley; *New York Times*, August 26, 1863.

41. Charles P. Stone to N. P. Banks, New Orleans, November 12, 1863, Charles P. Stone to Judge of the US District Court, New Orleans, November 12, 1863, Charles P. Stone to Rufus Waples, New Orleans, November 12, 1863, *OR*, series 1, vol. 26, pt. 1, 793–795.

42. Charles P. Stone to President [Lincoln], New Orleans, February 15, 1864, Abraham Lincoln Papers at the Library of Congress.

43. Charles P. Stone to N. P. Banks, Alexandria, March 23, 1864, *OR*, series 1, vol. 34, pt. 1, 178–179.

44. Anders, *Damp Sand*, 45; Ludwell H. Johnson, *The Red River Campaign: Politics and Cotton in the Civil War* (Kent, OH: Kent State University Press, 1993), 115–116; Gary Dillard Joiner, *One Damn Blunder from Beginning to End: The Red River Campaign of 1864* (Wilmington, DE: Scholarly Resources, Inc., 2003), 75–78; Hollandsworth, *Pretense*, 181.

45. *Report of the Joint Committee on the Conduct of the War at the Second Session of the Thirty-Eighth Congress* (Washington, DC: Government Printing Office, 1865), 190; Richard B. Irwin, *History of the Nineteenth Army Corps* (New York: G. P. Putnam's Sons, 1892), 427; Anders, *Damp Sand*, 92.

46. Anders, *Damp Sand*, 61–62; Moore, ed., *Rebellion Record*, vol. 8, 555, 562; C. S. Sargent to Major George B. Drake, Grand Ecore, April 12, 1864, William J. Landram to Captain Oscar Mohr, Grand Ecore, April 12, 1864, Brigadier General A. E. Lee to George B. Drake, New Orleans, April 29, 1864, *OR*, series 1, vol. 34, pt. 1, 271, 290, 456.

47. Anders, *Damp Sand*, 66–67.

48. Alphonse le Duc to George B. Drake, Grand Ecore, April 11, 1864, Report of Colonel William Shaw, April 15, 1864, *OR*, series 1, vol. 34, pt. 1, 270, 355; George B. Drake to Lieutenant Colonel Richard B. Irwin, Grand Ecore, April 11, 1864, *OR*, series 1, vol. 24, pt. 3, 128.

49. Joiner, *Blunder*, 147; Johnson, *Red River*, 206; Hollandsworth, *Pretense*, 198.

50. Field Order No. 21, Grand Ecore, April 16, 1864, *OR*, series 1, vol. 34, pt. 3, 175; Johnson, *Red River*, 219; Winkler, *Scapegoats*, 55–56.

51. Garrison, *Strange Battles*, 123–124; Anders, *Damp Sand*, 92; Charles P. Stone Appointments, Commissions and Promotions File, Adjutant General's Office Records, RG94, National Archives, cited in Holein, *Ball's Bluff*, 133.

52. C. P. Stone to Adjutant General, Cairo, June 29, 1864, Charles P. Stone to Adjutant General, Cairo, July 11, 1864, Charles P. Stone Appointments, Commissions and Promotions File, Adjutant General's Office Records, RG 94, National Archives, cited in Holein, *Ball's Bluff*, 134–135.

53. Marvel, *Mr. Lincoln*, 279; Edwin M. Stanton to Lieutenant General Grant, Washington, DC, April 20, 1864, *OR*, series 1, vol. 34, pt. 3, 234; Calendar, August 9, 1864, Simon, ed., *Personal Papers of Ulysses S. Grant*, vol. 11, 456; Special Orders

270, Headquarters of the Army, August 13, 1864; Special Orders No. 46, Headquarters, Second Division, Fifth Corps, August 21, 1864, *OR*, series 1, vol. 42, pt. 1, 143, 342.

54. George W. Cullum, *Biographical Register of the Officers and Graduates of the U. S. Military Academy at West Point*, 3rd Ed., vol. 2 (Boston: Houghton, Mifflin and Company, 1891), 218; C. P. Stone to Lieutenant Colonel F. T. Locke, Headquarters, Second Division, September 2, 1864, *OR*, series 1, vol. 42, pt. 2, 665–666; Marvel, *Mr. Lincoln*, 279; Holein, *Ball's Bluff*, 135.

CHAPTER TEN: STONE PASHA

1. Porter, *Stone*, 20; Charles P. Stone Certificate of Citizenship and Oath of Allegiance, General Records, Department of State, Passport Applications, 1795–1925, RG 59, National Archives.
2. Porter, *Stone*, 20; Sara Yorke Stevenson, *Maximilian In Mexico: A Woman's Reminiscences of the French Intervention, 1862–1867* (New York: Century Company, 1899), 174.
3. Temple Bayliss, "Soft Coal and Hard Times: An Account of the Dover Mines and Other Coal Mines that Changed Goochland," *Goochland County Historical Society Magazine*, 37/38 (2005–2006): 27–29; Cullum, *Biographical Register* 2, 218.
4. Bayliss, "Soft Coal," 27–29.
5. Ibid.; *The Central Water-Line from the Ohio River to the Virginia Capes, Connecting the Kanawha and James Rivers, Affording the Shortest Outlet of Navigation from the Mississippi Basin to the Atlantic* (Richmond, VA: Gary, Clemmitt & Jones, Printers, 1869), 93–96.
6. Porter, Stone, 20.
7. Lyslee Meyer, *The Farther Frontier: Six Case Studies of Americans and Africa, 1848–1936* (Selinsgrove: Susquehanna University Press, 1992), 66; William B. Hesseltine and Hazel C. Wolf, *The Blue and Gray on the Nile* (Chicago: University of Chicago Press, 1961), 37.
8. W. W. Loring, *A Confederate Soldier in Egypt* (New York: Dodd, Mead & Company, Publishers, 1884), 77.
9. Charles P. Stone to George Bancroft, Cairo, February 20, 1873, George Bancroft Papers, Massachusetts Historical Society, Boston, MA.
10. James Morris Morgan, *Recollections of a Rebel Reefer* (Boston: Houghton Mifflin, 1917), 296.
11. Hesseltine and Wolf, *Blue and Gray*, 52.
12. Ibid., 101.
13. Ibid., 101–102.
14. De Leon, Edwin, *The Khedive's Egypt: Or the Old House of Bondage Under New Masters* (London: Sampson Low, Marston, Searle & Rivington, 1877), 337–339.
15. John P. Dunn, *Khedive Ismail's Army* (London: Routledge, 2005), 59; Hesseltine and Wolf, *Blue and Gray*, 77.
16. Morgan, *Recollections*, 302.
17. Loring, *Confederate*, 349, 356–357; Dunn, *Army*, 59.
18. Dunn, *Army*, 48–49; Hesseltine and Wolf, *Blue and Gray*, 87–89.
19. Charles P. Stone to George Bancroft, Cairo, July 21, 1873, Bancroft Papers, Massachusetts Historical Society.

20. Hesseltine and Wolf, *Blue and Gray*, 81, 86; Dunn, *Army*, 49.

21. William McE. Dye, *Moslem Egypt and Christian Abyssinia: Or Military Service Under the Khedive, In His Provinces and Beyond their Borders, As Experienced by the American Staff* (New York: Atkin & Prout, Publishers, 1880), 74; Dunn, *Army*, 51–52; Hesseltine and Wolf, *Blue and Gray*, 81.

22. Charles P. Stone, *Asuntos Militares en Egipto* (Havana: Tipografia de "El Eco Militar," 1884), 31; Andrew McGregor, *A Military History of Modern Egypt: From the Ottoman Conquest to the Ramadan War* (Westport, CT: Praeger Security International, 2006), 141.

23. Dye, *Moslem Egypt*, 74–75.

24. De Leon, *Khedive's Egypt*, 361.

25. De Leon, *Khedive's Egypt*, 429–432; R. E. Colston, "Stone Pasha's Work in Geography," *Journal of the American Geographical Society of New York* 19 (1887): 48.

26. Charles Chaille-Long, *My Life on Four Continents* (London: Hutchinson and Co., 1912), 66.

27. Ibid., 136, 157–158; McGregor, *Military History*, 142.

28. Hesseltine and Wolf, *Blue and Gray*, 60–63.

29. Pierre Crabites, *Americans in the Egyptian Army* (London: George Routledge & Sons, 1938), 17–18.

30. Michael B. Oren, *Power, Faith and Fantasy: America in the Middle East, 1776 to the Present* (New York: W. W. Norton & Company, 2007), 234–235; Stephen W. Sears, *George B. McClellan: The Young Napoleon* (New York: Ticknor & Fields, 1988), 393.

31. John Russell Young, *Around the World with General Grant: A Narrative of the Visit of General US Grant, Ex-President of the United States to Various Countries in Europe, Asia, and Africa in 1877, 1878, 1879*, vol. 1 (New York: American News Company, 1879), 234; Jesse Root Grant and Henry Francis Granger, *In the Days of My Father, General Grant* (New York: Harper and Brothers, 1925), 267; John Y. Simon, ed., *The Personal Memoirs of Julia Dent Grant* (Carbondale: Southern Illinois University Press, 1975), 221.

32. Elbert Farman, *Along the Nile: An Account of the Visit to Egypt of General Ulysses S. Grant and His Tour Through that Country* (New York: Grafton Press, 1904), 31–32; James D. McCabe, *A Tour Around the World by General Grant* (New York: National Publishing Company, 1879), 295.

33. J. F. Packard, *Grant's Tour Around the World* (Philadelphia: H. W. Kelley, 1880), 230–231.

34. Cassandra Vivian, *Americans in Egypt, 1770–1915: Explorers, Consuls, Travelers, Soldiers, Missionaries, Writers, and Scientists* (Jefferson, NC: McFarland & Company, Inc., Publishers, 2012), 171, 178; Young, *Around the World*, 239; Simon, ed., *Julia Dent Grant*, 229.

35. Simon, ed., *Julia Dent Grant*, 263.

36. Hesseltine and Wolf, *Blue and Gray*, 114–115.

37. Dunn, *Army*, 114.

38. Ibid., 99, 107.

39. Ibid., 104.

40. Hesseltine and Wolf, *Blue and Gray*, 180–182; McGregor, *Military History*, 145.

41. Dunn, *Army*, 110–111; McGregor, *Military History*, 144; *Proceedings of the Royal Geographical Society* 20 (London: Savile Row, 1876), 382.

42. Hesseltine and Wolf, *Blue and Gray*, 178-179; McGregor, *Military History*, 146–147.

43. Dunn, *Army*, 108.

44. Hesseltine and Wolf, *Blue and Gray*, 183–184, 186; Dunn, *Army*, 134.

45. Hesseltine and Wolf, *Blue and Gray*, 184–186, 257.

46. Dunn, *Army*, 141–148. The American version of the campaign, the Battle of Gura, and its aftermath can be found in Loring, *Confederate*, chapters 10 and 11, and Dye, *Moslem Egypt*, chapters 30–50.

47. Patrick Richard Carstens, *The Encyclopedia of Egypt During the Reign of the Mehemet Ali Dynasty, 1798–1952; The People, Places, and Events that Shaped Nineteenth-Century Egypt and Its Sphere of Influence* (Victoria, BC: Friesen Press, 2014), 313–314; Dunn, *Army*, 148–149; Hesseltine and Wolf, *Blue and Gray*, 213.

48. Hesseltine and Wolf, *Blue and Gray*, 233–234; Carstens, *Encyclopedia*, 51; Elbert Farman, *Egypt and Its Betrayal: An Account of the Country During the Periods of Ismail and Tewfik Pashas, and How England Acquired a New Empire* (New York: Grafton Press, 1908), 154–155.

49. Hesseltine and Wolf, *Blue and Gray*, 244–245.

50. Henry M. Field, *On the Desert: With a Brief Review of Recent Events in Egypt* (New York: Charles Scribner's Sons, 1883), 5; George M. Wheeler, *Report Upon the Third Geographical Conference and Exhibition at Venice, Italy, 1881* (Washington, DC: Government Printing Office, 1885), 43.

51. William N. Armstrong, *Around the World with a King* (New York: Frederick A. Stokes Company, Publishers, 1904), 181–182.

52. Ibid., 183.

53. Ibid., 184.

54. Ibid., 190.

55. Crabites, *Americans*, 262–263; Efraim Karsh and Inari Karsh, *Empires in the Sand: The Struggle for Mastery in the Middle East, 1789–1923* (Cambridge, MA: Harvard University Press, 1993), 51; Hesseltine and Wolf, *Blue and Gray*, 245.

56. Hesseltine and Wolf, *Blue and Gray*, 246; Michael Barthorp, *War on the Nile: Britain, Egypt, and the Sudan, 1882–1898* (Poole, Dorset, UK: Blandford Press, 1984), 28.

57. Oren, *Power*, 264.

58. Field, *Desert*, 16.

59. Ibid.; Barthorp, *Nile*, 31.

60. Charles P. Stone, "The Bombardment of Alexandria: Rejoinder by Stone Pasha," *Century* 28 (October 1884): 954; Fanny Stone, "Diary of an American Girl in Cairo During the War of 1882," *Century* 28 (June 1884): 289.

61. Vivian, *Americans*, 183.

62. Oren, *Power*, 263; Stone, "Diary," 290; Chaille-Long, *My Life*, 272–273.

63. Donald Featherstone, *Tel-el-Kebir, 1882: Wolseley's Conquest of Egypt* (London: Osprey Military, 1993), 11–12, 18–20; Stone, "Diary," 291; Stone, "Bombardment," 953; Oren, *Power*, 266–267.

64. Charles P. Stone, "Stone-Pacha and the Secret Despatch," *Journal of the Military Service Institution of the United States* 8 (1887): 94–95; Stone, "Diary," 290, 292; Ibid., 291.

65. Stone, "Diary," 290, 292.
66. Ibid., 291
67. Ibid., 292, 293, 295.
68. Ibid., 293.
69. Ibid., 296.
70. Ibid., 297.
71. Ibid.
72. Ibid., 298, 300–301.
73. Ibid., 301–302.
74. Field, *Desert*, 11–12.
75. Hesseltine and Wolf, *Blue and Gray*, 249.

CHAPTER ELEVEN: LIBERTY'S ENGINEER

1. *Johnson's Revised Universal Cyclopedia: A Scientific and Popular Treasury of Useful Knowledge*, vol. 7 (New York: A. J. Johnson & Co., 1886), 187.
2. Elizabeth Mitchell, *Liberty's Torch: The Great Adventure to Build the Statue of Liberty* (New York: Atlantic Monthly Press, 2014), 157; *New York Times*, February 12, 1885; *Army and Navy Journal* 20 (April 28, 1883): 882.
3. Edward Berenson, *The Statue of Liberty: A Transatlantic Story* (New Haven: Yale University Press, 2012), 20–22.
4. *Army and Navy Journal* 20 (May 26, 1883): 981; *New York Times*, October 24, 1883; Yasmin Sabina Kahn, *Enlightening the World: The Creation of the Statue of Liberty* (Ithaca: Cornell University Press, 2010), 169; Cara A. Sutherland, *The Statue of Liberty* (New York: Barnes & Noble Books, 2003), 48; Mitchell, *Torch*, 167, 181; Barry Moreno, *The Statue of Liberty Encyclopedia* (New York: Simon & Schuster, 2000), 157.
5. Stone, "Diary," 288.
6. Stone, "Bombardment," 953–956.
7. Kahn, *Enlightening*, 155–157; Mitchell, *Torch*, 147–150, 154.
8. *New York Times*, August 6, 1884; Mitchell, *Torch*, 181–182.
9. *New York Times*, August 6, 1884; Berenson, *Transatlantic*, 78, 80, 85–86; Bernard A. Weisberger, Christian Blanchet and Bertrand Dard, *Statue of Liberty: The First Hundred Years* (New York: American Heritage, 1985), 85.
10. *New York Times*, August 6, 1884; Weisberger, *Hundred*, 85.
11. *New York Times*, October 24, 1883, January 30, 1884; Charles P. Stone, "The Navigation of the Nile," *Science* 4 (November 14, 1884): 456–457; Weisberger, *Hundred*, 79–80; Sutherland, *Statue*, 54.
12. Weisberger, *Hundred*, 90; Sutherland, *Statue*, 50–51, 55; Mitchell, *Torch*, 197–198.
13. Mitchell, *Torch*, 202.
14. Ibid., 198–200.
15. Ibid., 207–209.
16. Ibid., 217.
17. Ibid., 216; Andre Gschaedler, *True Light on the Statue of Liberty and Its Creator* (Narberth, PA: Livingston Publishing Company, 1966), 128.
18. *New York Times*, March 3, 1886; Mitchell, *Torch*, 227–228.

19. *New York Times,* October 6, 15, 1886.
20. Mitchell, *Torch,* 231.
21. John J. Garnett, *Official Programme, The Statue of Liberty: Its Conception, Its Construction, Its Inauguration* (New York: D. W. Dinsmore & Co., 1886), 61–62.
22. Garnett, *Programme,* 63–64, 66, 70; Mitchell, *Torch,* 239–240.
23. New York, *The Sun,* October 29, 1886; Garnett, *Programme,* 70–77; Mitchell, *Torch,* 242.
24. Mitchell, *Torch,* 249–250; Kahn, *Enlightening,* 179.
25. *New York Times,* October 30, 1886.
26. *New York Times,* January 1, 1887.
27. Weisberger, *Hundred,* 132, 137.
28. New York, *The Sun,* January 25, 1887; *New York Tribune,* January 25, 1887; Mitchell, *Torch,* 260; Thomas Stone Croft, *The Stone Family of East Feliciana Parish Louisiana and Their Ancestry* (n.p.: Recaptured Moments, 2008), 274.
29. *New York Times,* January 28, 1887.

EPILOGUE

1. *New York Times,* February 10, 1887; Mitchell, *Torch,* 261; Porter, Stone, 22; Mexican War Pension, Affidavit of Witness, October 27, 1887, Mrs. Jeannie Stone, Mexican War Pension Files, Records of the Veterans Administration, RG 15, National Archives; Chapter 386, 50th Congress, 1st Session, June 9, 1888, *Statutes at Large of the United States of America from December 1887 to March 1889* (Washington, DC: Government Printing Office, 1890), 1080.

BIBLIOGRAPHY

MANUSCRIPT COLLECTIONS

Library of Congress, Washington, DC.

Bancroft, George. Papers. Massachusetts Historical Society, Boston, MA.

Certificate of Incorporation, San Francisco and Sacramento Rail Road Company. Incorporation Papers, California State Archives, Sacramento.

Dispatches from United States Consuls in Guaymas, Mexico, vol. 1, November 27, 1832–December 31, 1867. RG 59. Records of the Department of State, National Archives, Washington, DC.

Dispatches from United States Ministers to Mexico, 1823–1906. RG 59. Records of the Department of State, National Archives, Washington, DC.

Documents for the History of California: The Papers of Thomas O. Larkin. Vol. 9. Part 3. Numbers 379 and 427. Bancroft Library, University of California, Berkeley.

Generals' Papers, Entry 159, Stone, Charles P. RG 94. Records of the Adjutant General's Office, National Archives, Washington, DC.

Gorgas, Josiah. "Journal of a Campaign in Mexico." Josiah and Amelia Gorgas Family Papers. W. S. Hoole Special Collections Library. University of Alabama, Birmingham.

Huger, Benjamin. War Journal of an Ordnance Chief: Being the Mexican War Diary of Captain Benjamin Huger, Chief of Ordnance, Army of Invasion. Typescript edited by Jeffrey L. Rhoades. National Museum of American History, Washington, DC.

Journal of Orders, Benicia Arsenal Log. Benicia Historical Museum Collection, Benicia, CA.

Letters Received by the Adjutant General (Main Series), 1822–1860. RG 94. Records of the Adjutant General's Office, National Archives, Washington, DC.

Mexican War Pension Files, Stone, Jeannie. RG 15. Records of the

Veterans Administration, National Archives, Washington, DC.

Ord, Edward Otho Cresap. Papers. Bancroft Library, University of California, Berkeley.

Passport Applications, 1795–1925. RG 59. Records of the Department of State, National Archives, Washington, DC.

Returns from United States Military Posts, 1800–1916, Ordnance Depot Near Benicia, 1851–1854. RG 94. Records of the Adjutant General's Office, National Archives, Washington, DC.

Returns from United States Military Posts, 1800–1916, West Point, New York, 1845–1846. RG 94. Records of the Adjutant General's Office, National Archives, Washington, DC.

Stone, Charles P. to Benson Lossing. Letter, November 5, 1866. William L. Clements Library, University of Michigan, Ann Arbor.

United States Military Academy Appointment File, Stone, Charles P. RG 94. Records of the Adjutant General's Office, National Archives, Washington, DC.

Abraham Lincoln Papers at the Vital Records from the *Daily Evening Bulletin*, San Francisco, California, 1856, 1857. Compiled by the Puerta de Oro Chapter, Genealogical Records Committee, California State Society of the Daughters of the American Revolution, 1943. Typescript. California Room, California State Library, Sacramento.

Whiting Family Letters, New York Historical Society Museum and Library, New York.

PUBLISHED MANUSCRIPT COLLECTIONS

The Collected Works of Abraham Lincoln. Vol. 5. Edited by Roy P. Basler. New Brunswick: Rutgers University Press, 1953.

The Papers of Ulysses S. Grant, Volume 1, 1837–1861. Edited by John Y. Simon. Carbondale: Southern Illinois University Press, 1967.

The Works of James Buchanan, Comprising His Speeches, State Papers, and Private Correspondence. Vol. 10. Edited by John Bassett Moore. New York: Antiquarian Press Ltd., 1960.

GOVERNMENT DOCUMENTS

Annual Report of the Board of Regents of the Smithsonian Institution, Showing Operations, Expenditures and Condition of the Institution for the Year 1860. Washington, DC: George W. Bowman, Printer, 1860.

Annual Report of the Quartermaster and Adjutant-General. California State Assembly Document Number 6. Session of 1856.

Correspondence Between the Late Secretary of War and General Wool. House Executive Document. 35th Congress. 1st Session. Serial Volume 956. Report 88. 1857–1858.

Hewett, Janet B., et al. *Supplement to the Official Records of the Union and Confederate Armies.* Volumes 1, 4, and 10. Wilmington, NC: Broadfoot Publishing Company, 1994.

Instructions and Correspondence Between the Government and Major General Wool, In Regard to His Operations on the Coast of the Pacific, January 3, 1855. Senate Executive Document Number 16. 33rd Congress. 2nd Session. Serial Volume 751.

Message from the President of the United States, in Answer to a Resolution of the Senate Calling for Further Information in Relation to the Formation of State Government in California; and Also in Relation to the Condition of Civil Affairs in Oregon. Executive Document Number 52. 31st Congress. 1st Session.

Message from the President of the United States to the Two Houses of Congress at the Commencement of the First Session of the Thirty-Sixth Congress. Senate Executive Document Number 2. 36th Congress. 1st Session. Volume 1. Washington, DC: George W. Bowman, Printer, 1860.

Murray, John and James V Swantek. *1813–1997, The Watervliet Arsenal: A Chronicle of the Nation's Oldest Arsenal.* Watervliet, NY: Watervliet Arsenal, 1998.

Official Register of the Officers and Cadets of the US Military Academy, West Point, New York. June, 1842–June, 1845.

Regulations for the Government of the Ordnance Department. Washington, DC: Printed by Gideon, 1852.

"Report of the Colonel of Ordnance." Message from the President of the United States to the Two Houses of Congress at the Commencement of the Second Session of the Thirty-Second Congress, December 6, 1852. House Executive Document Number 1. Part 2. 32nd Congress. 2nd Session. Serial Volume 674.

Report of the Secretary of War. With Statements Showing the Contracts Made Under Authority of that Department During the Year 1851. Senate Executive Document Number 12. 32nd Congress. 1st Session.

Report of the Joint Committee on the Conduct of the War. 37th Congress. 3rd Session. Washington, DC: Government Printing Office, 1863.

Report of the Joint Committee on the Conduct of the War. 38th Congress. 2nd Session. Washington, DC: Government Printing Office, 1865.

Resolution Concerning Brigadier General Charles P. Stone. Senate. 37th Congress. 2nd Session. April 21, 1862. *The Congressional Globe.* 1862. Part 2.

Revised Regulations for the Army of the United States, 1861. Philadelphia: J. B. Lippincott & Co., Publishers, 1862.

Senator McDougall Speaking for the Resolution Concerning Brigadier General Charles P. Stone. 37th Congress. 2nd Session. April 15, 1862. *The Congressional Globe.* 1862. Part 2.

Senator Wade Speaking Against the Resolution Concerning Brigadier General Charles P. Stone. Senate. 37th Congress. 2nd Session. April 15, 1862. *The Congressional Globe.* 1862. Part 2.

Statutes at Large of the United States of America from December, 1887 to March, 1889. Chapter 386. 50th Congress. 1st Session. June 9, 1888. Washington, DC: Government Printing Office, 1890.

The War of the Rebellion: A Compilation of the Official Records of the Union and Confederate Armies. Washington, DC: Government Printing Office, 1881–1901.

BOOKS AND PAMPHLETS

Acuna, Rodolfo F. *Sonoran Strongman: Ignacio Pesqueira and His Times.* Tucson: University of Arizona Press, 1974.

Akelyn, John. *Biographical Dictionary of the Confederacy.* Westport, CT: Greenwood Press, 1977.

Alcaraz, Ramon, Barreiro, Alejo, Payno, Manuel, Prieto, Guillermo, Ramirez, Ignacio, et al. *Apuntes para la historia de la Guerra entre Mexico y los Estados Unidos.* Mexico, 1848.

Almada, Francisco R. *Diccionario de Historia, Geografia y Biografia Sonorenses.* Chihuahua, 1952.

Anders, Curt. *Disaster in Damp Sand: The Red River Expedition.* Carmel: Guild Press of Indiana, 1997.

Anderson, Robert. *An Artillery Officer in the Mexican War, 1846–1847.* New York: G. P. Putnam's Sons, 1911.

Armstrong, William N. *Around the World with a King.* New York: Frederick A. Stokes Company, Publishers, 1904.

Bancroft, Hubert Howe. *History of California, Volume 1, 1542–1800.* San Francisco: History Company, Publishers, 1886.

Bandel, Eugene. *Frontier Life in the Army, 1854–1861.* Glendale, CA: Arthur H. Clark Company, 1932.

Barthorp, Michael. *War on the Nile: Britain, Egypt and the Sudan, 1882–1898.* Poole, Dorset, UK: Blandford Press, 1984.

Bauer, K. Jack. *The Mexican War, 1846–1848.* Lincoln: University of Nebraska Press, 1992.

Beatie, Russell H. *Army of the Potomac: Birth of Command, November 1860–September 1861.* Boston: Da Capo Press, 2002.

Beatie, Russell H. *Army of the Potomac: McClellan Takes Command, September 1861–February 1862.* Boston: Da Capo Press, 2004.

Before the Joint Commission of the United States and Mexico In the Matter of the Claim of the Lower California Company, Memorial and Exhibits (Iturbide Grant). New York: Evening Post Steam Presses, 1870.

Before the Joint Commission of the United States and Mexico In the Matter of the Claim of the Lower California Company, Memorial and Exhibits (Sonora Survey). New York: Evening Post Steam Presses, 1870.

Berenson, Edward. *The Statue of Liberty: A Transatlantic Story*. New Haven: Yale University Press, 2012.

Blaine, James G. *Twenty Years of Congress: From Lincoln to Garfield, with a Review of the Events which Led to the Political Revolution of 1860*. Norwich, CT: Henry Bill Publishing Company, 1884.

Blair, Harry and Tarshis, Rebecca. *Lincoln's Constant Ally: The Life of Colonel Edward D. Baker*. Portland: Oregon Historical Society, 1960.

Box, Michael James. *Capt. James Box's Adventures and Explorations in New and Old Mexico*. New York: Derby & Jackson, Publishers, 1861.

Brooks, N. C. *A Complete History of the Mexican War: Its Causes and Consequences, Comprising an Account of the Various Military and Naval Operations from Its Commencement to the Treaty of Peace*. Philadelphia: Gregg, Elliot & Co., 1851.

Bruegmann, Robert. *Benicia: Portrait of an Early California Town, An Architectural History*. Fairfield, CA: James Stevenson, Publisher, 1997.

Butler, Stephen. *A Documentary History of the Mexican War*. Richardson, TX: Descendants of Mexican War Veterans, 1995.

California Pioneer Register and Index, 1524–1848: Including Inhabitants of California, 1769–1800, and a List of Pioneers Extracted from the History of California by Hubert Howe Bancroft. Baltimore: Regional Publishing Company, 1964.

Calvo Berber, Laureano. *Nociones de Historia de Sonora*. Mexico, DF: Libreria de Miguel Porrua, SA, 1958.

Cantu, Gaston Garcia. *Las Invasiones Norteamericnas en Mexico*. Mexico, DF: Fondo Cultura Econnomica, 1996.

Carstens, Patrick Richard. *The Encyclopedia of Egypt During the Reign of the Mehemet Ali Dynasty, 1798–1952: The People, Places and Events that Shaped Nineteenth Century Egypt and its Sphere of Influence*. Victoria, BC: Friesen Press, 2014.

The Central Water-Line from the Ohio River to the Virginia Capes, Connecting the Kanawha and James Rivers, Affording the Shortest Outlet of Navigation from the Mississippi Basin to the Atlantic. Richmond, VA: Gary, Clemmitt & Jones, Printers, 1869.

Chaille-Long, Charles M. *My Life on Four Continents*. London: Hutchinson and Co., 1912.

Chapin, Charles Wells. *Sketches of the Inhabitants of Old Springfield, and Other Citizens of the Present Century, and Historic Mansions of "Ye Olden Tyme."* Springfield, MA: Springfield Printing and Binding Company, 1893.

Clarke, Dwight L. *William Tecumseh Sherman: Gold Rush Banker.* San Francisco: California Historical Society, 1969.

Clary, David A. *Eagles and Empire: The United States, Mexico and the Struggle for a Continent.* New York: Bantam Books, 2009.

Coffman, Edward M. *The Old Army: A Portrait of the American Army in Peacetime, 1784–1898.* New York: Oxford University Press, 1986.

Colville, Samuel. *Colville's San Francisco Directory, 1856, 1857.* Vol. 1. San Francisco: Commercial Steam Presses, Monson, Valentine & Co., 1856.

Conkling, Alfred R. *The Life and Letters of Roscoe Conkling: Orator, Statesman, Advocate.* New York: Charles L. Webster & Company, 1889.

Corti, Count Egon Caesar. *Maximilian and Charlotte of Mexico.* Vol. 1. New York: Alfred A. Knopf, 1928.

Cowell, Josephine W. *History of the Benicia Arsenal, Benicia, California, January 1851–December 1962.* Berkeley, CA: Howell-North Books, 1963.

Crabites, Pierre. *Americans in the Egyptian Army.* London: George Routledge & Sons, 1938.

Croffut, William, Editor. *Fifty Years in Camp and Field: Diary of Major General Ethan Allen Hitchcock, USA.* New York: G. Putnam's Sons, 1909.

Croft, Thomas Stone. *The Stone Family of East Feliciana Parish, Louisiana, and Their Ancestry.* NP: Recaptured Moments, 2008.

Cross, Ira. *Financing an Empire: History of Banking in California.* Vol. 1. Chicago: S. J. Clarke Publishing Company, 1927.

Cullum, George W. *Biographical Register of the Officers and Graduates of the US Military Academy at West Point, Volume 2, 1841–1847.* New York: D. Van Nostrand, 1868.

Cullum, George W. *Biographical Register of the Officers and Graduates of the US Military Academy at West Point.* 3rd Edition. Vol. 2. Boston: Houghton, Mifflin and Company, 1891.

Cunningham, Edward. *The Port Hudson Campaign, 1862–1863.* Baton Rouge: Louisiana State University Press, 1963.

Cushing, Caleb. *Contract of the Mexican Government for the Survey of the Public Lands of the State of Sonora.* Washington, DC: 1860.

Cutler, Carl C. *Queens of the Western Ocean: The Story of America's Mail and Passenger Sailing Lines.* Annapolis, MD: United States Naval Institute, 1961.

DeLeon, Edwin. *The Khedive's Egypt: Or the Old House of Bondage Under New Masters.* London: Sampson, Low, Searle & Rivington, 1877.

Detzer, David. *Dissonance: The Turbulent Days Between Fort Sumter and Bull Run.* Orlando, FL: Harcourt Inc., 2006.

Diccionario Porrua de Historia, Biografia y Geografia de Mexico, Segunda Edicion. Mexico, DF: Editorial Porrua, SA, 1965.

Dillon, Richard, *Great Expectations: The Story of Benicia, California.* Benicia: Benicia Heritage Book Inc., 1980.

Duganne, A. J. H. *Camps and Prisons: Twenty Months in the Department of the Gulf.* New York: J. P. Robens, Publisher, 1865.

Dunn, John P. *Khedive Ismail's Army.* London: Routledge, 2005.

Dupuy, Trevor, Johnson, Kurt and Bongard, David L. *The Harper Encyclopedia of Military Biography.* New York: HarperCollins, 1992.

Dye, William McE. *Moslem Egypt and Christian Abyssinia: Or Service Under the Khedive, In His Provinces and Beyond their Borders, As Experienced by the American Staff.* New York: Atkin & Prout, Publishers, 1880.

Eisenhower, John S. D. *Agent of Destiny: The Life and Times of Winfield Scott.* New York: Free Press, 1997.

Farman, Elbert. *Along the Nile: An Account of the Visit to Egypt of General Ulysses S. Grant and His Tour Through that Country.* New York: Grafton Press, 1904.

Farman, Elbert. *Egypt and Its Betrayal: An Account of the Country During the Periods of Ismail and Tewfik Pashas, and How England Acquired a New Empire.* New York: Grafton Press, 1904.

Farwell, Byron. *Ball's Bluff: A Small Battle and its Long Shadow.* McLean, VA: EPM Publications Inc., 1990.

Featherstone, Donald. *Tel-el-Kebir, 1882: Wolseley's Conquest of Egypt.* London: Osprey Military, 1993.

Field, Henry M. *On the Desert: With a Brief Review of Recent Events in Egypt.* New York: Charles Scribner's Sons, 1883.

Field, Ron. *The Mexican-American War.* London: Brassey UK Ltd., 1997.

Fishel, Edwin C. *The Secret War for the Union: The Untold Story of Military Intelligence in the Civil War.* Boston: Houghton Mifflin Company, 1996.

Ford, Andrew E. *History of the Fifteenth Massachusetts Volunteer Infantry in the Civil War, 1861–1864.* Clinton: Press of W. J. Coulter, 1898.

Frazer, Robert W. Editor. *Mansfield on the Condition of the Western Forts, 1853–1854.* Norman: University of Oklahoma Press, 1963.

Frost, John. *Pictorial History of Mexico and the Mexican War: Comprising an Account of the Ancient Aztec Empire, the Conquest by Cortes, Mexico Under the Spaniards, the Mexican Revolution, the Republic, the Texan War, and the Recent War with the United States.* Philadelphia: Charles Desilver, 1852.

Furgurson, Ernest B. *Freedom Rising: Washington in the Civil War.* New York: Alfred A. Knopf, 2004.

Garrett, John J. *Official Programme, The Statue of Liberty: Its Conception, Its Construction, Its Inauguration.* New York: Dinsmore & Co., 1886.

Garrison, Webb, Jr. *Strange Battles of the Civil War.* New York: Bristol Park Books, 2001.

General Scott and His Staff; Comprising Memoirs of Generals Scott, Twiggs, Smith, Lane, Cadwalader, Patterson and Pierce; Colonels Childs, Riley, Harney and Butler, and Other Distinguished Officers Attached to General Scott's Army; Together with Notices of General Kearny, Colonel Doniphan, Colonel Fremont, and Other Officers Distinguished in the Conquest of California and New Mexico. Philadelphia: Gregg, Elliot & Co., 1848.

Goodwin, Doris Kearns. *Team of Rivals: The Political Genius of Abraham Lincoln*. New York: Simon & Schuster, 2005.

Grant, Jesse Root and Granger, Henry Francis. *In the Days of My Father, General Grant*. New York: Harper and Brothers, 1925.

Grant, Julia Dent. *The Personal Memoirs of Julia Dent Grant*. John Y. Simon, Editor. Carbondale: Southern Illinois University Press, 1975.

Grant, Ulysses S. *Personal Memoirs of US Grant*. 2nd Edition. Vol. 1. New York: DeVinne Press, 1895.

Gschaedler, Andre. *True Light on the Statue of Liberty and its Creator*. Narberth, PA: Livingston Publishing Company, 1966.

Hague, Harlan. *Thomas O. Larkin: A Life of Patriotism and Profit in Old California*. Norman: University of Oklahoma Press, 1990.

Harrington, Fred Harvey. *Fighting Politician: Major General N. P. Banks*. Philadelphia: University of Pennsylvania Press, 1948.

Haydon, F. Stansbury. *Military Ballooning During the Early Civil War*. 2000 Edition. Baltimore: Johns Hopkins University Press, 1941.

Hesseltine, William B. and Wolf, Hazel C. *The Blue and Gray on the Nile*. Chicago: University of Chicago Press, 1961.

Hittell, John S. *The Resources of California: Comprising Agriculture, Mining, Geography, Commerce &c, and the Past and Future Development of the State*. San Francisco: A. Roman and Company, 1863.

Holien, Kim Bernard. *The Battle at Ball's Bluff*. Orange, VA: Publisher's Press, 1995.

Hollandsworth, James. *Pretense of Glory: The Life of Nathaniel P. Banks*. Baton Rouge: Louisiana State University Press, 1998.

Howard, William F. *The Battle of Ball's Bluff: "The Leesburg Affair."* Lynchburg, VA: H. E. Howard, 1994.

Hunton, Eppa. *Autobiography of Eppa Hunton*. Richmond, VA: William Byrd Press, 1933.

Hussey, George A. *History of the Ninth Regiment, NYSM—NGSNY (Eighty-Third NY Volunteers), 1845–1888*. New York: Veterans of the Regiment, 1888.

Irwin, Richard B. *History of the Nineteenth Army Corps*. New York: G. P. Putnam's Sons, 1892.

Jackson, William Turrentine. *Wagon Roads West: A Study of Federal Road Surveys and Construction in the Trans-Mississippi West, 1846–1869*. New Haven: Yale University Press, 1965.

Jaques, John Wesley. *Three Years' Campaign of the Ninth NYSM, During the Southern Rebellion.* New York: Hilton & Co., Publishers, 1865.

Jenkins, John S. *History of the War Between the United States and Mexico from the Commencement of Hostilities to the Ratification of the Treaty of Peace.* Auburn, NY: Derby and Miller, 1851.

Jermann, Donald R. *Civil War Battlefield Orders Gone Awry: The Written Word and Its Consequences in Thirteen Engagements.* Jefferson, NC: McFarland & Company, 2012.

Johnson, Ludwell H. *The Red River Campaign: Politics and Cotton in the Civil War.* Kent, OH: Kent State University Press, 1993.

Johnson, Timothy. *A Gallant Little Army: The Mexico City Campaign.* Lawrence: University Press of Kansas, 2007.

Johnson's Revised Universal Cyclopedia: A Scientific and Popular Treasury of Useful Knowledge. Vol. 1. New York: A. J. Johnson & Co., 1886.

Joiner, Gary Dillard. *One Damn Blunder from Beginning to End: The Red River Campaign of 1864.* Wilmington, DE: Scholarly Resources Inc., 2003.

Kahn, Yasmin Sabina. *Enlightening the World: The Creation of the Statue of Liberty.* Ithaca: Cornell University Press, 2010.

Karsh, Efraim and Karsh, Inari. *Empires in the Sand: The Struggle for Mastery in the Middle East, 1789–1923.* Cambridge: Harvard University Press, 1993.

Kline, Michael J. *The Baltimore Plot: The First Conspiracy to Assassinate Abraham Lincoln.* Yardley, PA: Westholme Publishing, 2008.

Leech, Margaret. *Reveille in Washington, 1860–1865.* New York: Harper & Brothers, 1941.

Lewis, Lloyd. *Captain Sam Grant.* Boston: Little, Brown and Company, 1950.

Libura, Krystyna, Moreno, Luis Gerardo Morales and Marquez, Jose Velasco. *Echoes of the Mexican-American War.* Toronto: Groundwood Books, 2004.

Lockwood, John and Charles Lockwood. *The Siege of Washington: The Untold Story of Twelve Days that Shook the Union.* New York: Oxford University Press, 2011.

Long, David F. *Gold Braid and Foreign Relations: Diplomatic Activities of US Naval Officers, 1798–1898.* Annapolis, MD: Naval Institute Press, 1988.

Loring, W. W. *A Confederate Soldier in Egypt.* New York: Dodd, Mead & Company, Publishers, 1884.

Lossing, Benson J. *Pictorial History of the Civil War of the United States of America.* Vol. 1. Philadelphia: George W. Childs, Publisher, 1866.

Mansfield, Edward D. *The Life and Services of General Winfield Scott, Including the Siege of Vera Cruz, the Battle of Cerro Gordo, and the Battles in the Valley of Mexico, to the Conclusion of Peace and his Return to the United States*. New York: A. S. Barnes & Co., 1852.

Maria de Bustamante, Carlos. *El Nuevo Bernal Diaz del Castillo, o sea historia de la invasion de los angloamericanos en Mexico*. Mexico, DF: Conacluta, 1990.

Martinez, Pablo L. *A History of Lower California*. Translated by Ethel Duffy Turner. Mexico, DF: Editorial Baja California, 1960.

Marvel, William. *Mr. Lincoln Goes to War*. Boston: Houghton Mifflin, 2006.

Marzalek, John F. *Commander of All Lincoln's Armies: A Life of General Henry W. Halleck*. Cambridge, MA: Belknap Press of Harvard University Press, 2004.

Mayo, John. *Commerce and Contraband on Mexico's West Coast in the Era of Barron, Forbes & Co., 1821–1859*. New York: Peter Lang, 2006.

McCabe, James D. *A Tour Around the World by General Grant*. New York: National Press Publishing Company, 1879.

McCaffery, James M. Editor. *Surrounded by Dangers of All Kinds: The Mexican War Letters of Lieutenant Theodore Laidley*. Denton: University of North Texas Press, 1997.

McClellan, George B. *McClellan's Own Story: The War for the Union, The Soldiers Who Fought It, The Civilians Who Directed It, and His Relations to It and Them*. New York: Charles L. Webster & Company, 1887.

McGregor, Andrew. *A Military History of Egypt: From the Ottoman Conquest to the Ramadan War*. Westport, CT: Praeger Security International, 2006.

Meyer, Lyslee. *The Farther Frontier: Six Case Studies of Americans and Africa, 1848–1936*. Selinsgrove, PA: Susquehanna University Press, 1992.

Miller, Robert Royal. Editor. *The Mexican War Journal and Letters of Ralph W. Kirkham*. College Station: Texas A&M University Press, 1991.

Mitchell, Elizabeth. *Liberty's Torch: The Great Adventure to Build the Statue of Liberty*. New York: Atlantic Monthly Press, 2014.

Moore, Frank. Editor. *Rebellion Record: A Diary of American Events, With Documents, Narratives, Illustrative Incidents, Poetry, Etc.* Vol. 4. Documents. New York: G. Putnam, 1862.

Moreno, Barry. *Statue of Liberty Encyclopedia*. New York: Simon & Schuster, 2000.

Morgan, James A., III. *A Little Short of Boats: The Fights at Ball's Bluff and Edward's Ferry, October 21–22, 1861*. Fort Mitchell, KY: Ironclad Publishing Inc., 2004.

Morgan, James Morris. *Recollections of a Rebel Reefer*. Boston: Houghton Mifflin Company, 1917.

Mowry, Sylvester. *Arizona and Sonora: The Geography, History and Resources of the Silver Region of North America*. 3rd Edition. New York: Harper & Brothers, 1864.

Moyano, Angela Pahissa. *California y sus Relaciones con Baja California.* Mexico, DF: Fondo del Cultural Economica, 1983.

Oren, Michael B. *Power, Faith and Fantasy: America in the Middle East, 1776 to the Present.* New York: W. W. Norton & Company, 2007.

Oswandel, J. Jacob. *Notes on the Mexican War, 1846–47–48.* Philadelphia, 1885.

Packard, J. F. *Grant's Tour Around the World.* Philadelphia: H. W. Kelley, 1880.

Patterson, Charles J. *The Military Heroes of the War With Mexico: With a Narrative of the War.* Philadelphia: William A. Leary & Company, 1850.

Perret, Geoffrey. *Lincoln's War: The Untold Story of America's Greatest President as Commander in Chief.* New York: Random House, 2004.

Peskin, Allan. *Winfield Scott and the Profession of Arms.* Kent, OH: Kent State University Press, 2003.

Pfanz, Donald C. *Richard S. Ewell: A Soldier's Life.* Chapel Hill: University of North Carolina Press, 1998.

Poore, Ben Perley. *Perley's Reminiscence of Sixty Years in the National Metropolis.* Vol. 2. Philadelphia: Hubbard Brothers, 1888.

Rafuse, Ethan S. *McClellan's War: The Failure of Moderation in the Struggle for the Union.* Bloomington: Indiana University Press, 2005.

Reid, Brian Holden. *The Civil War and the Wars of the 19th Century.* New York: Harper Collins, 1999.

Rhoades, Jeffrey. *Scapegoat General: The Story of General Benjamin Huger, C. S. A.* Hamden, CT: Archon Books, 1985.

Ripley, R. S. *The War With Mexico.* Vol. 2. New York: Harper & Brothers, 1849.

Scott, Winfield. *Memoirs of Lieut. General Scott, LLD.* New York: Sheldon & Company, Publishers, 1864.

Scroggs, William O. *Filibusters and Financiers: Story of William Walker and His Associates.* New York: Macmillan Company, 1916.

Sears, Stephen W. Editor. *The Civil War Papers of George B. McClellan: Selected Correspondence, 1860–1865.* New York: Ticknor & Fields, 1989.

Sears, Stephen W. *Controversies and Commanders: Dispatches from the Army of the Potomac.* Boston: Houghton Mifflin, 1999.

Sears, Stephen W. *George B. McClellan: The Young Napoleon.* New York: Ticknor & Fields, 1988.

Semmes, Raphael. *Service Afloat and Ashore During the Mexican War.* Cincinnati: Wm. H. Moore & Co., Publishers, 1851.

Sherman, William T. *Memoirs of General William T. Sherman.* Vol. 1. 2nd Edition. New York: D. Appleton and Company, 1904.

Shuck, Oscar T. *Bench and Bar in California: History, Anecdotes, Reminiscences.* San Francisco: Occident Printing House, 1889.

Simpson, Brooks D. *Ulysses S. Grant: Triumph Over Adversity, 1822–1865.* Boston: Houghton Mifflin, 2000.

Smith, Winston and Charles Judah. Editors. *Chronicles of the Gringos: The US Army in the Mexican War, 1846–1848, Accounts of Eyewitnesses and Combatants.* Albuquerque: University of New Mexico Press, 1968.

Speech of Hon. J. A. McDougall of California on the Arrest of Gen. Stone and the Rights of the Soldier and Citizen. Washington, DC: L. Towers & Co., 1862.

Speer, Lonnie R. *Portals to Hell: Military Prisons of the Civil War.* Mechanicsburg, PA: Stackpole Books, 1997.

Stevenson, Sara Yorke. *Maximilian in Mexico: A Woman's Reminiscences of the French Intervention, 1862–1867.* New York: Century Company, 1899.

Stone, Charles P. *Asuntos Militares en Egito.* Havana: Tipografia de "El Eco Militar," 1884.

Stone, Charles P. *Notes on the State of Sonora.* Washington, DC: Samuel Polkinghorn, Printer, 1860.

Stone, Charles P., Jecker, J. B. y ca., Isham, J. B. G. *A todos a quienes interesare.* Washington, DC, 1860.

Stout, Joseph A. *Schemers and Dreamers: Filibustering in Mexico, 1848–1921.* Fort Worth: Texas Christian University Press, 2002.

Sutherland, Cara A. *The Statue of Liberty.* New York: Barnes & Noble Books, 2003.

Tap, Bruce. *Over Lincoln's Shoulder: The Committee on the Conduct of the War.* Lawrence: University Press of Kansas, 1998.

Taylor, Alexander S. *A Historical Summary of Baja California from Its Discovery to 1867.* Pasadena, CA: Socio-Technical Books, 1971.

Title Papers of the Lower California Company, to Lands, Etc., in the Territory of Lower California and in the States of Sonora and Sinaloa of the Republic of Mexico. New York: Evening Post Steam Presses, 1870.

Townsend, E. D. *Anecdotes of the Civil War of the United States.* New York: D. Appleton and Company, 1884.

Valades, Adrian. *Historia de la Baja California, 1850–1880.* Mexico, DF: Universidad Nacional Autonoma de Mexico, 1974.

Vivian, Cassandra. *Americans in Egypt, 1770–1915: Explorers, Consuls, Travelers, Soldiers, Missionaries, Writers and Scientists.* Jefferson, NC: McFarland & Company Inc., Publishers, 2012.

Voss, Stuart F. *On the Periphery of Nineteenth Century Mexico: Sonora and Sinaloa, 1810–1877.* Tucson: University of Arizona Press, 1982.

Warren, T. Robinson. *Dust and Foam; or, Three Oceans and Two Continents.* New York: Charles Scribner, 1849.

Waugh, John C. *The Class of 1846, from West Point to Appomattox: Stonewall Jackson, George McClellan and their Brothers.* New York: Ballantine Books, 1994.

Weisberger, Bernard A., Blanchet, Christian, and Bertrand, David. *Statue of Liberty: The First Hundred Years*. New York: American Heritage, 1985.

Welcher, Frank D. *The Union Army, 1861–1865: Organization and Operations, Volume 1, The Eastern Theater*. Bloomington: Indiana University Press, 1989.

Wheeler, George M. *Report Upon the Third Geographical Conference and Exhibition at Venice, Italy, 1881*. Washington, DC: Government Printing Office, 1885.

Wilcox, Cadmus. *History of the Mexican War*. Washington, DC: Church News Publishing Company, 1892.

Williams, T. Harry. *Lincoln and the Radicals*. Madison: University of Wisconsin Press, 1965.

Winders, Richard Bruce. *Mr. Polk's Army: The American Military Experience in the Mexican-American War*. College Station: Texas A&M University Press, 1997.

Winkle, Kenneth J. *Lincoln's Citadel: The Civil War in Washington, DC*. New York: W. W. Norton & Company, 2013.

Winkler, H. Donald. *Civil War Goats and Scapegoats*. Nashville: Cumberland House, 2008.

Winters, John D. *The Civil War in Louisiana*. Baton Rouge: Louisiana State University Press, 1963.

Woodworth, Stephen E. *Sherman*. New York: Palgrave Macmillan, 2009.

Wyllys, Rufus K. *The French in Sonora, 1850–1854: The Story of French Adventurers from California into Mexico*. Berkeley: University of California Press, 1932.

Young, John Russell. *Around the World with General Grant: A Narrative of the Visit of General US Grant, Ex-President of the United States to Various Countries in Europe, Asia and Africa in 1877, 1878, 1879*. Vol. 1. New York: American News Company, 1879.

ARTICLES

"The Arsenal at Benicia." *California Farmer and Journal of Useful Sciences* 3. May 10, 1855.

Bayliss, Temple. "Soft Coal and Hard Times: An Account of the Dover Mines and Other Coal Mines that Changed Goochland." *Goochland County Historical Society Magazine* 37/38. 2005–2006.

"Biographical Notice of the Late Dr. Alpheus Stone." *Boston Medical and Surgical Journal* 45. October 15, 1851.

Clark, Paul C. and Edward M. Mosley. "D-Day, Veracruz, 1847: A Grand Design." *Joint Force Quarterly*. Winter 1995–96.

Colston, R. E. "Stone Pasha's Work in Geography." *Journal of the American Geographical Society of New York* 19. 1887.

Irwin, Richard B. "Ball's Bluff and the Arrest of General Stone." *Battles and Leaders of the Civil War.* Vol. 2. New York: Century Company, 1884–1887.

Kielbowicz, Richard B. "The Telegraph, Censorship and Politics at the Outset of the Civil War." *Civil War History* 40. June 1994.

Nicolay, John G. and John Hay. "The President Elect in Springfield." *Century* 35. November 1887.

Phillips, Matthew. "Bungled River Crossing." *Military History Magazine's Great Battles.* July 1991.

Porter, Fitz-John. "Charles Pomeroy Stone." *Eighteenth Annual Reunion of the Association of Graduates of the United States Military Academy at West Point, New York, June 9, 1887.* East Saginaw, MI: Evening News Printing and Binding House, 1887.

Poston, Charles D. "Building a State in Apache-Land." *Overland Monthly* 24. July 1894.

"The Powder Magazine at Benicia." *California Farmer and Journal of Useful Sciences* 3. April 26, 1855.

Ripley, R. S. "Chapultepec and the Garitas of Mexico." *Southern Quarterly Review.* January 1853.

Snyder, Timothy R. "Securing the Potomac: Colonel Charles P. Stone and the Rockville Expedition, June–July 1861." *Catoctin History.* Issue no. 11. 2009.

Stone, Charles P. "The Bombardment of Alexandria: Rejoinder by Stone Pasha." *Century* 28. October 1884.

Stone, Charles P. "A Dinner with General Scott." *Magazine of American History* 11. January–June 1884.

Stone, Charles P. "The Navigation of the Nile." *Science* 4. November 14, 1884.

Stone, Charles P. "Stone-Pacha and the Secret Despatch." *Journal of the Military Service Institution of the United States* 8. 1887.

Stone, Charles P. "Washington in March and April 1861." *Magazine of American History* 14. July 1885.

Stone, Charles P. "Washington on the Eve of the War." *Battles and Leaders of the Civil War.* Vol. 1. New York: Century Company, 1884–1887.

Stone, Fanny. "Diary of an American Girl in Cairo During the War of 1882." *The Century* 28. June 1884.

Trefousse, Hans L. "The Joint Committee on the Conduct of the War: A Reassessment." *Civil War History* 10. March 1964.

Vandiver, Frank E. "The Mexican War Experience of Josiah Gorgas." *Journal of Southern History* 13. August 1947.

"A Visit to Popocaptetl." *Putnam's Monthly Magazine* 1. April 1853.

Williams, T. Harry. "Investigation, 1862." *American Heritage* 6. December 1954.

NEWSPAPERS

Army and Navy Journal
Los Angeles Star.
New York. *Evening Post.*
New York. *The Sun.*
New York Times.
New York Tribune.
San Diego Herald.
San Francisco. *Daily Alta California.*
San Francisco. *Daily Evening Bulletin.*
Tubac. *Weekly Arizonian.*
Ures, Sonora. *La Estrella de Occidente*
Washington. *Evening Star.*

ONLINE SOURCES

Aztec Club of 1847 Website. http://www.walika.com/aztec/bios/stone.
"Biographical Memoir of Charles Henry Davis, 1807–1877, Read Before the National Academy, April 1896." National Academy of Sciences Biographical Memoirs. http://www.nasonline.org/pubications/biographi-calmemoirs.

INDEX

Devens, Charles, 123, 124, 125–126,
126–127, 129, 134
relieved of duty, 132
as Stone's pallbearer, 223
District of Columbia, 83
District of Columbia Militia, 120–121
Division of the Army of the
Potomac, 118
Dover Company, 180–181
Downieville, California, 54, 55
Drum, Simon, 29
Dutton, William, 7
Dye, William McEntyre, 187, 196

E. Remington and Sons, 187
Edward's Ferry, 110, 112, 113, 114,
119, 122, 123, 124, 125, 126, 127,
128, 129, 131, 134–135
McClellan arrived at, 139
Stone rode back to, 136
Egypt:
Abyssinia and, 193–198
American officer corps in, 182
European influence in, 197–198,
199, 202
European occupation of, 201
internal strife in, 202
Stone's employment in, 181–182
Egyptian army, 194
Eiffel, Gustave, 215
Eightieth Article of War, 162–163
8th Virginia Infantry, 113, 120
Electoral College, 91
Ellsworth, Elmer, 109
El Penon, 27
*Encyclopedia of Egypt During the Reign
of the Mehemet Ali Dynasty*, ix
Engineer Corps, 9
Escandon, Antonio, 61
Evans, Nathan, 126, 131
Evarts, William, 221
Evening Bulletin, 79–80
Evening Post, 217
Ewell, Richard S., 77
Executive Square, 97, 99, 102, 103

Farragut, David, 6
Ferik Pasha, 183
Field Order Number 21, 177
15th Massachusetts Infantry, 119,
122, 124, 125, 126, 129, 130
50th Massachusetts Infantry, 170

1st Brigade, 178
1st California Infantry, 120, 127, 128,
129, 132
1st Minnesota Infantry, 119, 138
1st New Hampshire Infantry, 110
1st New York Infantry, 133
Florida Ship Canal Company,
211–212
Floyd, John B., 62–63, 77, 87
Folsom, Joseph, 52
Forest City, California, 54, 55
Fort Buchanan, 70, 74–75, 77
Fort Gura, 196
Fort Hamilton, 158, 159, 160
Stone's imprisonment at, 161–165,
166
Fort Lafayette, 157, 158, 159
Fort Point, 50
Fortress Monroe, 10, 12, 13, 14, 38,
40
Fort Sumter, 96, 97
Fort Vancouver, 47
42nd New York Infantry, 119, 128,
133
Fort Yuma, California, 40, 63
14th Infantry Regiment, 108–109,
157
Franklin, William B., 105, 106, 174
Fugitive Slave Act of 1850, 121

Garczynski, Edward Rudolf, 217
General Order Number Four, 103
General Post Office, 99, 102
Gilmore, G. A., 218
Gold Rush, x, 42–43, 52, 57, 60
Goodrich, Caspar F., 214
Gordon, Charles, 188–189, 194
Gorgas, Josiah, 14–15, 17
Gorman, Willis, 119, 122–123, 129,
132, 134
Gorman's brigade, 125, 138
Gosport Navy Yard, 101
Grant, Frederick, 191
Grant, Julia, 191–192
Grant, Ulysses S., 6, 47–48, 52, 175
in Egypt, 191–192
in Mexico City, 32
Red River campaign and, 177
Statue of Liberty and, 212
Guadalupe Hidalgo, 31, 34
Guaymas, 66, 68, 69, 70–71, 72, 76,
77